International Conference on Environmental Engineering and Green Technology

IOP Conference Series: Earth and Environmental Science
Volume 1057

Online
4-6 April 2022

Editors:

P Bhuvaneswari
Abyot Asres
Abebaw Abizo
Wegayehu Enbeyle

ISBN: 978-1-7138-6020-4
ISSN: 1755-1307

Printed from e-media with permission by:

Curran Associates, Inc.
57 Morehouse Lane
Red Hook, NY 12571

Some format issues inherent in the e-media version may also appear in this print version.

Institute of Physics
Dirac House, Temple Back
Bristol BS1 6BE UK

Phone: 44 1 17 929 7481
Fax: 44 1 17 920 0979

techtracking@iop.org

Additional copies of this publication are available from:

Curran Associates, Inc.
57 Morehouse Lane
Red Hook, NY 12571 USA
Phone: 845-758-0400
Fax: 845-758-2633
Email: curran@proceedings.com
Web: www.proceedings.com

TABLE OF CONTENTS

IOP Conf. Series: Earth and Environmental Science 1057 (2022) 011001 doi:10.1088/1755-1315/1057/1/011001

PREFACE

1st International Conference on Environmental Engineering and Green Technology -ICEEGT'22 held during April 5 & 6, 2022. ICEEGT'22 is a meeting place for Environmental Engineering and Green Technology specialists from academia and industry.

The main purpose of the conference is to promote innovative research in environmental engineering and green technology. In addition, researchers, developers, engineers, students, and practitioners will be better able to share scientific knowledge. The annual conference will focus on environmental science and development, as well as other connected themes.

Yazhli Global Multidisciplinary Research Organization (YGMRO), Tamilnadu, India, and Mizan-Tepi University, Ethiopia, are co-hosting the conference. The summit was intended to provide a scholarly forum for the exploration of environmental issues, sustainable development, environmental food & agricultural and green technology.

We appreciate the hard work of the organizing committee and the reviewers who worked diligently to ensure that all of the papers presented at the conference met rigorous scientific standards. In light of this, we'd like to thank our teammates for their assistance and advice throughout the editing process. Additionally, we'd like to thank Feichi from the IOP Conference Series office for all of his help and support.

The articles in this issue are meant to educate and inspire readers about environmental topics such as sustainable materials, environmental healthcare and safety, ecological engineering, pollution, environmental food & agriculture and green technology.

Chief Patron

S Vishwanathan, Chief Executive Officer, Yazhli Global Multidisciplinary Research Organization, India

Dr Ahmed Mustofa, President of Mizan -Tepi University, Ethiopia

Patron

Dr Kasahun Mulatu, Vice President for Academic Affairs, Mizan-Tepi University, Ethiopia

Dr Algene Malte De Guzman, Director of Internationalization, YGMRO & West Bay College, Philippines

Chair

Yideg Mamo, Tepi Campus Chief Executive Director Office, Mizan-Tepi University, Ethiopia

IOP Conf. Series: Earth and Environmental Science 1057 (2022) 011001 doi:10.1088/1755-1315/1057/1/011001

Andualem Henok, Teaching Hospital Campus Chief Executive Director Office, Mizan-Tepi University, Ethiopia

Editors
Dr P Bhuvaneswari, Research Head, Yazhli Global Multidisciplinary Research Organization, Tirupur, Tamilnadu, India

Dr Abyot Asres, Vice President, Research and Community Service Development, Mizan -Tepi University, Ethiopia

Dr Abebaw Abizo, Director, Research Directorate, Mizan-Tepi University, Ethiopia

Wegayehu Enbeyle, Director,International Relations and Partnership, Mizan -Tepi University, Ethiopia

Steering Committee

Dr D R V A Sharath Kumar, Associate Professor, ECE Department, MVSR Engineering College, Hyderabad

Dr Defekadu Haile, Department of Plant Science, Mizan-Tepi University, Ethiopia

Bizuwork Derebew, Dean, College of Natural and Computational Science, Mizan-Tepi University, Ethiopia

Wendmagegn Addis, Tepi Campus, Academic Programs Directorate, Mizan-Tepi University, Ethiopia

Sintayehu Gashew, Director, Research ,Community Services, industry Linkage and Technology Transfer Directorate Tepi Campus, Mizan-Tepi University, Ethiopia

Wondwosen Wondmu, Dean, College of Agriculture, Mizan-Tepi University, Ethiopia

Mengistu Ketema, Dean, College of Engineering and Technology, Mizan-Tepi University, Ethiopia

Semgnew Assemie, Dean, School of Computing and Informatics, Mizan-Tepi University, Ethiopia

Speakers

Dr.Algene Malte De Guzman, Director of Internationalization, YGMRO & West Bay College, Philippines

Dr Nalini Kanta Sahoo, Principal, SRM Modinagar College of Pharmacy, SRM Institute of Technology

IOP Conf. Series: Earth and Environmental Science 1057 (2022) 011001 doi:10.1088/1755-1315/1057/1/011001

Dr. Tesfaye Gezahegn, Post Graduate Directorate Director, Mizan Tepi University

ICEEGT'22 conference took place virtually and the following information are:

· Why the meeting is held/was held virtually (COVID-19/travel restrictions, etc).

The meeting was held virtually because of Covid-19 travel restrictions.

· Location: where the organisers will be/were.

Organizers: Yazhli Global Multidisciplinary Research Organization (YGMRO), Tamilnadu, India, and Mizan-Tepi University, Ethiopia.

· The Model: e.g., the plenary, the contributions, the time spent to deliver the talks by each speaker and interactivity (e.g. Q&A sessions)

The tasks that we would perform on the day of conference would be

1) **Welcoming Address [Started at IST 11.30]**
2) **President Speech**
3) **Vice President Speech**
4) **Director of Internationalization Speech**
5) **YGMRO -MD Speech**
6) **Chief Guest**
7) **Keynote Speaker – Session 1 (April 5)**
8) **Paper Presentation (8 papers each 15 minutes, Q&A -5 minutes per paper)**
9) **Keynote Speaker – Session 2 (April 5)**
10) **Paper Presentation (8 papers each 15 minutes, Q&A -5 minutes per paper)**
11) **Keynote Speaker – Session 3 (April 6)**
12) **Paper Presentation (7 papers each 15 minutes, Q&A -5 minutes per paper)**
13) **Best Paper -Results Announcement [ICCEGT_62, 66,78]**
14) **Vote of Thanks.**

· The participants: location and the attendance number expected.

The participants are from Vietnam, Ecuador, Poland, Slovakia & India. Attendance expected on

· The technology: What technology did you use, including any technical difficulties.

Virtual Event Platform: CISCO Webex

Meeting link:

https://meet210.webex.com/meet210/j.php?MTID=mfc35a6d2de08e01cab531c9bea32d740

Meeting number:

2640 993 2104

Join from a video or application

Dial 26409932104@meet210.webex.com

You can also dial 210.4.202.4 and enter your meeting number.

Join by phone

Use VoIP only

· Drawbacks in the delivery of the conference virtually (if any), e.g., impact on the community.

Went well without any issues

Peer Review Statement

All papers published in this volume have been reviewed through processes administered by the Editors. Reviews were conducted by expert referees to the professional and scientific standards expected of a proceedings journal published by IOP Publishing.

- **Type of peer review:** Double Anonymous
- **Conference submission management system:** Morressier
- **Number of submissions received:** 92
- **Number of submissions sent for review:** 25
- **Number of submissions accepted:** 23
- **Acceptance Rate (Submissions Accepted / Submissions Received × 100):** 25
- **Average number of reviews per paper:** 2
- **Total number of reviewers involved:** 29
- **Contact person for queries:**
  Name: Dr P Bhuvaneswari
  Email: p.bhuvaneswari@ygmro.com
  Affiliation: Research Head, Yazhli Global Multidisciplinary Research Organization, Tirupur, Tamilnadu, India

IOP Conf. Series: Earth and Environmental Science 1057 (2022) 012001 doi:10.1088/1755-1315/1057/1/012001

Thermal comfort conditions in traditional and sustainable, low - energy buildings

N Krawczyk[1] **and A Kapjor**[2]

[1]Kielce University of Techology, al. Tysiaclecia P.P.7, 25-314 Kielce, Poland
[2]University of Zilina, Univerzitná 8215, 010 26 Žilina, Slovakia

E-mail: nkrawczyk@tu.kielce.pl

Abstract. Nowadays, each of us spends more and more time indoors. The article focuses on research on thermal comfort in a traditional and sustainable building. A sustainable building is equipped with heat pumps, a wind turbine, solar collectors and photovoltaic cells. It also has mechanical ventilation, which helps to minimize energy consumption and lower carbon dioxide emissions. The research was carried out using two methods, anonymous questionnaires and the Testo 400 microclimate meter. Based on the conducted research, the thermal sensations of the subjects and their preferences as to the prevailing conditions in the examined buildings were determined. The study also made it possible to compare the PMV index calculated using the Fanger model with TSV (obtained on the basis of anonymous questionnaires).

1. Introduction

These days we spend more and more time indoors. Defining the adjusted parameters of the microclimate becomes a huge problem because each of us wants to feel good in the rooms. The issue of thermal comfort is an increasingly common topic. During learning or work, we want to be provided with adequate climatic conditions. Thermal sensations are a biased concept. It is impossible to create a place that will meet the needs of all people. Therefore, the solution is to create conditions that suit as many people as possible, so as to guarantee a state of indifference (one in which it is neither too cold nor too warm).

PMV (Predicted Mean Vote) is a function of physical activity, clothing thermal conductivity, air temperature and velocity, humidity, and mean radiant temperature. This index is reported on a seven-point scale, which was developed by O. Fanger in the 1970s based on current standards: ISO 7730 [1] together with PN-EN 16798-1:2019 [2]. -3 is cold and +3 is hot. The comfort range was assumed from -0.5 to +0.5. On the other hand, the PPD ratio for buildings of the 2nd category of public utility buildings) should not exceed 10%. Thus, the thermal comfort index is considered statistically important, for example, if we have a value of 0, it means that people feel most comfortable. Fanger also developed the PPD (Predicted Percentage of Dissatisfied) indicator, i.e. the predicted percentage of dissatisfied people with the conditions in the tested room.

Research on thermal comfort is still a very interesting issue among researchers. Research in modern buildings was also carried out by Majewski et al. [3]. They observed significant differences between the actual and forecast values of thermal comfort. The authors reached similar conclusions [4,5]. Research on thermal comfort in intelligent buildings was also carried out by Kolková et al. [6], in which they showed that the optimal temperature range in offices is from 23 to 27°C. In residential buildings, research on thermal comfort was carried out by [7]. The authors presented the differences between the Fanger model and subjective feelings. Their results showed that PMV in summer was below zero, and vice versa in winter. The respondents were satisfied with the air temperature of 21.5°C in apartments heated during the winter. On the other hand, in air-conditioned rooms in summer, the satisfactory air temperature for the respondents was 23°C. Other authors [8] focused on what has the greatest impact on thermal comfort. They proved that the most important factor influencing thermal comfort is air temperature. Vilcekova et al. [9] examined the air quality in schools attended by children aged 6 to 15, as well as 5 educators. The authors showed that the fundamental problem during the research was the noise level, low light level and high CO_2 concentration. Which significantly influenced the differences between PMV (Predicted Mean Vote) and TSV (Thermal Sensations Vote).bSimilar results were presented by the authors [10]. The authors [11] conducted research in an educational building on 41 students aged 17-22. Respondents were dissatisfied with the thermal conditions in the area with coffee shops and a lounge area. They proved that you can improve the thermal sensation by increasing the temperature by 1°C. Manu et al. [12] worked out studies of thermal comfort in 16 buildings in three seasons: winter, spring and summer. The research was carried out for buildings with natural ventilation, with air conditioning and a mixed mode. The authors explained that office residents are tolerant of high temperatures, and presented that the Fanger model inflates the feeling of warmer ones. Chai et al. [13] discussed the materials of thermoregulatory systems in terms of heat exchange between the inside and outside of the building. It is important to ensure thermal comfort for users and save energy at the same time. On the other hand, other authors [14] investigated the influence of clothing temperature on the comfort of users through the process of heat transfer. They showed that fabrics can improve the thermal comfort of users and thus it will have an impact on energy savings in rooms. Hesaraki et at. [15] discussed LTH / HTC radiant systems. This study showed that thanks to these systems, it is possible to save almost 30% of energy, and at the same time provide more favourable conditions of thermal comfort. Additionally, they revealed that a combination of two systems (radiant and air), which will also improve thermal comfort.

The article compares two buildings of the Kielce University of Technology, in traditional and sustainable low - energy buildings. This paper aims to provide information on thermal comfort conditions in a traditional and low - energy building. Attention should be paid to the environmental aspect. Sustainable low - energy buildings obtain their energy from renewable energy sources, making them self - sufficient. Sustainable buildings reduce the building's environmental impact by minimizing energy consumption for heating and air conditioning. They also ensure that the thermal environment is user - friendly, making modern buildings more environmentally friendly.

2. Material and method

In order to perform the research, measurements of microclimate parameters were carried out in two buildings of the Kielce University of Technology. The first building is a sustainable building and the second is a traditional building. During the performed microclimate tests for both buildings ventilation was turned off. The intelligent building has mechanical ventilation and BMS (Building Manager System) that manages the entire building. It uses renewable energy sources (photovoltaic panels, solar collectors and heat pumps) that make the building self-sufficient. The figures below show a sustainable building (fig. 1) and a traditional building (fig. 2).

A sustainable building has been designed with sustainability in mind. Due to very good thermal insulation properties, it is similar to the standard of passive construction [16]. The measurements were made using two methods. The Testo 400 microclimate meter was used for the first method. Testo 400 meter collected data on the microclimate conditions in the room from probes placed on a tripod. The unit was placed in the centre of the room, specifically at the level of the thermal centre of gravity.

Air temperature, air velocity, carbon dioxide concentration and relative air humidity for the intelligent building are 24.40°C, 0.1 m/s, 725 ppm and 29.11%, while for the traditional one: 27.50°C, 0.05 m/s, 580 ppm and 50.90% [17]. The second method was to fill in questionnaires by students. The questionnaire contained questions about the thermal impressions of the microclimate The results of the questionnaires made it possible to assess the thermal comfort of the rooms surveyed and their preferences as to the prevailing conditions.

Figure 1. Photo of the Energis building

Figure 2. Photo of the Traditional building

3. Results and discussion

The research in an intelligent building was carried out in January 2022, in which 14 people aged 21-22 participated, including 6 women and 8 men. The test results regarding the traditional building have been presented by the author in [19], which was carried out in June 2021. 20 people aged 24-30 participated in the research, including 3 women and 17 men. Figure 3 presents the percentage of the obtained responses regarding the thermal sensations in the rooms studied.

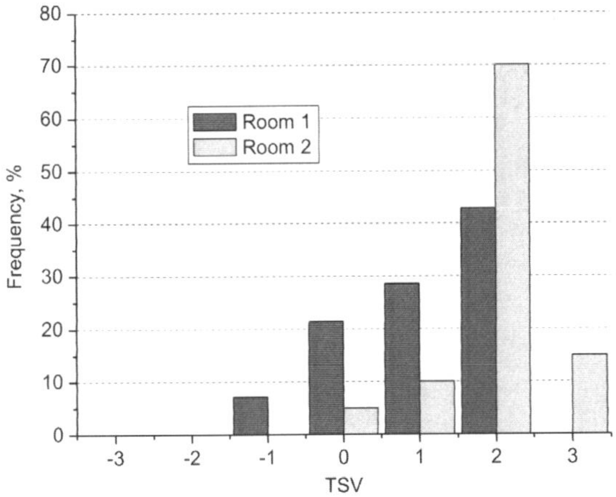

Figure 3. Frequency of responses given by the respondents regarding TSV (thermal sensations vote): room 1 - sustainable building, room 2 - traditional building

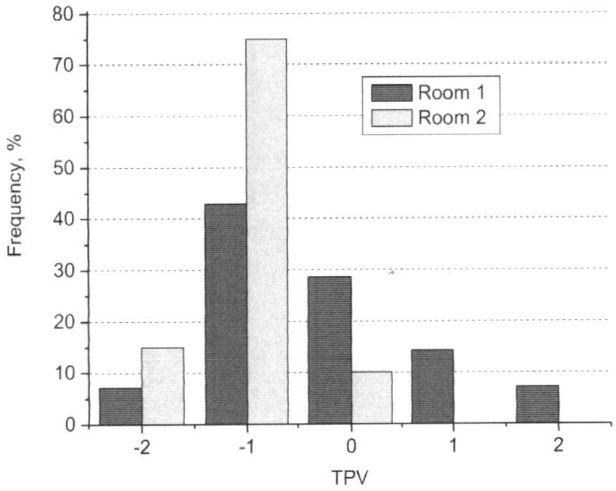

Figure 4. Frequency of responses given by the respondents regarding TPV (thermal preferences vote): room 1 - sustainable building, room 2 - traditional building

In figure 3, green is the percentage for a sustainable building, and yellow is for a traditional building. The respondents were asked about their thermal feelings, which are included on a seven-point scale. On the x-axis, these values mean: "-3" - too cold, "-2" - too cool, "-1" - pleasantly cool, "0" - comfortable, "1" - pleasantly warm, "2" - too warm, "3" - too hot. It can be seen that the better part of respondents chose the answer too warm. For room 1, this answer was 42.86%. The second most frequently chosen option for an intelligent building was the pleasantly warm answer. It amounted to 28.57% of all respondents. The answer comfortable was chosen by 3 out of 14 people, which is 21.43%. On the other hand, the pleasantly cool answer was the least chosen and constitutes 7.15%. For a traditional building it is as follows: respondents most often chose the answer too warm - 70%. The answer to too hot was 15%. The pleasantly warm answer was 10%, while the least chosen answer was comfortable, which was 5%. During the tests, ventilation was not used. It can be seen that the respondents did not feel any thermal comfort in the examined rooms. The next figure (Fig. 4) shows the expectations of the respondents with regard to air temperatures in the examined rooms. When asked about expectations regarding air temperatures, the respondents had the following answers to choose from: "-2" - definitely cooler, "-1" - cooler, "0" - unchanged, "1" - warmer, "2" - definitely warmer. For the first building, the most frequently given answer was the cooler. It amounted to 42.86%. The unchanged response is 28.57%, warmer - 14.29%, and definitely cooler and definitely warmer are at the same level (7.14%). In the second building, the most frequently chosen answer was the cooler, which was 75%. The least chosen option was unchanged (10%), while definitely cooler was 15%. The presented diagram shows that the respondents want the room to be cooler. The next figure (Fig. 5) shows the respondents' answers on the assessment of humidity.

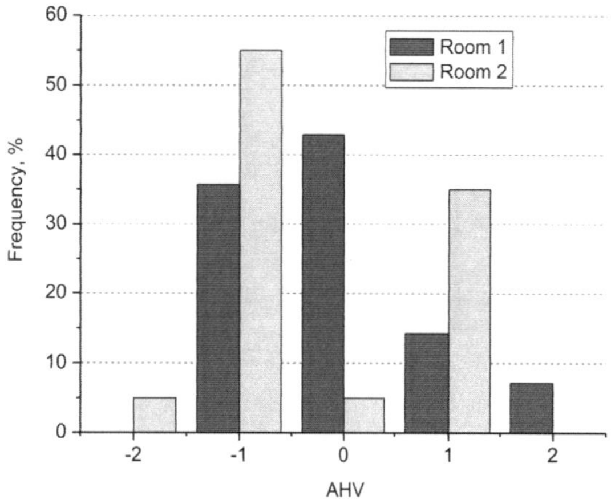

Figure 5. Frequency of the answers given by the respondents regarding AHV (air humidity vote): room 1 - sustainable building, room 2 - traditional building

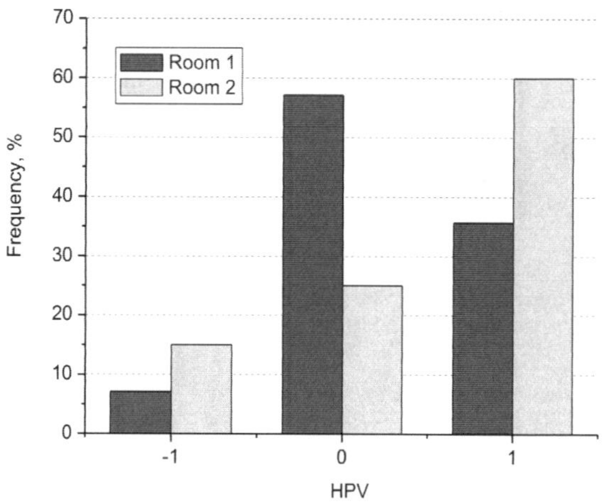

Figure 6. Frequency of answers given by the respondents regarding HPV (humidity preferences vote): room 1 - sustainable building, room 2 - traditional building

In the questionnaire, when asked about the assessment of air humidity in the examined rooms, the respondents had the following options: "-2" - too dry, "-1" - fairly dry, "0" - pleasantly, "1" - quite damp, " 2 "- too humid. For a sustainable building, the respondents most often chose the pleasantly options. This response was 42.86%. Another answer was fairly dry which is 35.71%. The least frequently chosen answer was too humid and it amounts to 7.14, and the quite damp option - 14.29%. In the first room, the humidity was at the level of 29.11% (the test was carried out in winter). However, for the second room, the humidity was 50.9%. The survey was carried out in the summer and it can be observed that the vast majority of respondents assessed the humidity as fairly dry, which is 55%. 35% of respondents chose quite damp, and too dry and pleasantly at the same level (5%). In the research, respondents were also asked about their preferences regarding humidity, as shown in Figure 6. The respondents, when asked about their preferences, had the following answers to choose from: "-1" - more dry, "0" - no change, "1" - more humid. For room 1, the respondents do not want changes to the air humidity in the tested room. The no change response was 57.14%. And for room 1, the same answer is 25%. The respondents in the traditional room want changes. 60% of respondents want the humidity in the room to be more humid. On the other hand, 5 respondents from the intelligent room chose this answer - 35.71%. In both buildings, the omitted answer was - more humid, which is 7.14% for room 1, 15% for room 2. The study also discussed the comparison of PMV calculated from the Fanger model and TSV obtained from the questionnaires, as shown in Figure 7.

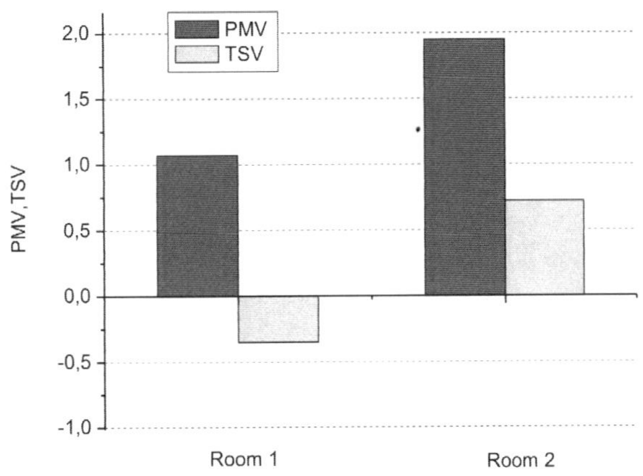

Figure 7. Comparison of TSV obtained using questionnaires and PMV calculated on the basis of the Fanger model for intelligent and traditional buildings: room 1 - sustainable building, room 2 - traditional building

Chart 7 presents the confrontation of the results of the average assessment of thermal comfort calculated from the Fanger model with the results obtained on the basis of the conducted surveys. The results of PMV are marked in green, and TSV in yellow. Based on the above chart, it can be concluded that the results from the Fanger model do not coincide with the results based on the questionnaires. These differences are: 1.42 for an intelligent building, and 1.23 for a traditional one. The range of thermal comfort is from -0.5 to +0.5. Such a discrepancy in the results may be due, inter alia, to the fact that other parameters (e.g. carbon dioxide concentration) were not taken into account in the Fanger model.

4. Conclusions

It can be seen from the presented results that the respondents assessed their thermal feelings as too warm. The vast majority of respondents want the indoor temperature to be cooler. When looking at air humidity, the respondents assessed sustainable building as pleasantly and they do not want changes. In contrast, in the traditional building, the respondents rated the relative humidity as fairy dry and 60% of the respondents want the humidity in the room to be more humid.

When analysing the results for the average assessment of thermal comfort calculated from the Fanger model with the results obtained on the basis of the surveys, significant differences between these values were found. It should be mentioned that the respondents did not feel comfortable in the examined rooms. This proves that the Fanger model is insufficient and requires modification. Over 42.86% of all respondents for the intelligent building were dissatisfied, and for the traditional building - over 85%. You can see significant differences for a sustainable and traditional building. More than a half more respondents (for a traditional building) are dissatisfied with the climatic conditions. A sustainable building creates more favourable microclimatic conditions. Thanks to the use of renewable energy sources, the building is self-sufficient. These conclusions allow further research to be continued more widely.

5. References

[1] ISO International Organisation for Standardization, Ergonomics of the thermal environment – Analytical determination and interpretation of thermal comfort using calculation of the PMV and PPD indices and local thermal comfort criteria, International Standard ISO 7730, 2005.

[2] PN-EN 16798-1:2019, Energy Performance of Buildings-Ventilation for Buildings-Part 1: Indoor Environmental Input Parameters for Design and Assessment of Energy Performance of Buildings Addressing Indoor Air Quality, Thermal Environment, Lighting and Acustics, 2019.

[3] Majewski G, Telejko M and Orman Ł J 2017 Preliminary results of thermal comfort analysis in selected buildings *Proc. Conf. on Interdisciplinatry Problems in Environmental Protection and Engineering EKO-DOK* (Boguszow-Gorce, Poland) E3S Web of Conferences vol. 17, 00056. DOI: 10.1051/e3sconf/20171700056

[4] Dębska L 2021 Assessment of the indoor environment in the intelligent building *Civil and Environmental Engineering* **17** 572 – 582 DOI: 10.2478/cee-2021-0058

[5] Krawczyk N and Krakowiak J 2021 The comparison of thermal comfort test results in selected traditional and modern buildings *Proc 10th Int. Conf. on Thermal Equipments, Renewable Energy and Rural Development* (Romania) E3S Web of Conferences vol. 286. DOI: 10.1051/e3sconf/202128602008

[6] Kolková Z, Hrabovský P, Florková Z and Lenhard R 2020 Analysis of ensuring thermal comfort using an intelligent control system *Proc. of XXII. Int. Conf. The Application of Experimental and Numerical Methods in Fluid Mechanics and Energy* (Slovakia) MATEC Web of Conferences vol. 328, 03017. DOI:10.1051/matecconf/202032803017

[7] Becker R, Paciuk M 2009 Thermal comfort in residential buildings – Failure to predict be Standard model BaE 44 948 - 960 DOI: 10.1016/j.buildenv.2008.06.011

[8] Jazizadeh F, Marin F M, Becerik-Gerber B 2013 A thermal preference scale for personalized comfort profile identification via participatory sensing BaE **68** 1440-149 DOI: 10.1016/j.buildenv.2013.06.011

[9] Vilcekova S, Meciarova L, Burdova E K, Katunska J, Kosicanova D, Doroudiani S 2017 Indoor environmental quality of classrooms and occupants' comfort in a special education school in Slovak Republic BaE **120** 29 - 40. DOI: 10.1016/j.buildenv.2017.05.001

[10] Mors S T, Hensen J L M, Loomans M G L C, Boerstra A C 2011 Adaptive thermal comfort in primary school classrooms: Creating and validating PMV-based comfort charts BaE **46** 2454 -2461 DOI: 10.1016/j.buildenv.2011.05.025

[11] Merabtine A, Maalouf C, Waheed Hawila A A, Martaj N and Polidori G 2018 Building energy audit, thermal comfort, and IAQ assessment of a school building: A case study *Building and Environment* **145** 62–76 DOI: 10.1016/j.buildenv.2018.09.015

[12] Manu S, Shukla Y, Rawal R, Thomas L E, de Dear R 2016 Field studies of thermal comfort across multiple climate zones for the subcontinent: India Model for Adaptive Comfort (IMAC) BaE **98** 55-70. DOI: 10.1016/j.buildenv.2015.12.019

[13] Chai J, Fan J 2022 Advanced thermal regulating materials and systems for energy saving and thermal comfort in buildings Materials Today Energy **24**, 100925 DOI: 10.1016/j.mtener.2021.100925

[14] Ma Z, Zhao D, Wang F, Yang R 2022 A novel thermal comfort and energy saving evaluation model for radiative cooling and heating textiles EaB **258**, 111842 DOI: 10.1016/j.enbuild.2022.111842

[15] Hesaraki A, Huda N 2022 A comparative review on the application of radiant low-temperature heating and high-temperature cooling for energy, thermal comfort, indoor air quality, design and control SETaA vol. 49 DOI: 10.1016/j.seta.2021.101661

[16] Piotrowski J Z, Orman Ł J, Lucas X, Zender – Świercz E, Telejko M and Koruba D 2014 Tests of thermal resistance of simulated walls with the reflective insulation *Proc. Int. Conf. Experimental Fluid Mechanics 2013* (Kutná Hora, Czech Republic) EPJ Web of Conferences vol. **67**, 02095. DOI: 10.1051/epjconf/20146702095

[17] Krawczyk N 2022 Comparison of thermal comfort in a traditional and intelligent building *Proc. Int. Conf. International Conference on Energy and Green Computing (Maroko)* E3S Web of Conferences vol. **336**. DOI: 10.1051/e3sconf/202233600019

Using Community Project in Teaching Food Security: A Student-centered Approach

Trang Nguyen[1] and Danh Vu[2,*]
[1]Institute of Biotechnology and Food Technology, Industrial University of Ho Chi Minh City, Vietnam
[2]Institute of Applied Technology, Thu Dau Mot University, Binh Duong, Vietnam
*Email: danhvc@tdmu.edu.vn

Abstract. This study aimed to implement community action-based project to help students apply knowledge of food security to the real world. The project had students involved in choosing a food security related topic, collecting data from a community, creating a data-driven action that is beneficial to the selected community, and reporting their work. To aid educators in successfully adopting a project-based learning approach in food security at undergraduate level, this manuscript explained the course framework, demonstrated students' works, discussed benefits and challenges when employing the community project in the course. The approach described here is applicable in other multidisciplinary fields.

1. Introduction

Sustainability includes sustainable food system that provides consumers safe and nutritious food in quantity and quality aspects, economically uses natural resources, and supports food security [1, 2]. This is a subject that requires cross-cutting knowledge and have high impact to the society. When introduced into the undergraduate program in Vietnam, food security is an emerging science course. The course taught at our institute focuses community challenge issues in ensuring food safety and nutrition when the world population reaches more than 9 billion people in 2050. The understand of how to solve food security problems highlights the need for students that they should be equipped knowledge about food security, but also provided opportunities to apply the knowledge in the real-world practices. Students will feel engaged when they experience of how the knowledge of food security can directly affect their lives and communities. However, the traditional teaching methods within classroom makes student less feel engaged in applying the learned knowledge of food security meaningfully in real world [3]. In Vietnam, education method mainly is lecture-based teaching, leading to ongoing issue in engaging students to scientific topics. Several studies reported that project-based learning (PBL) have potential to enhance student understanding of scientific content when inquiry is added to their curriculum [4-8]. In addition, the PBL supports student actively involve into real-world practices by using student's investigation, collaboration, social interactions, and knowledge sharing in the learning process and thus it helps students achieve their learning goals [9].

According to Blumenfeld et al., the PBL learning provides students opportunities to solve real world challenges by asking questions, designing experiments and conduct investigations, collecting, analyzing, interpreting data, drawing conclusions, and reporting findings [10]. Similarly, working on this PBL-based community project requires students' scientific inquiry, including 1) asking driving questions (related to food security issues), 2) choosing a community that may have beneficials from the research, 3) designing and planning steps to collect data that are essential to the community issues, 4) collecting and analyzing data, 5) making data-driven actions that is beneficial to the selected community, 6) discussing findings and reflection. In addition, students are encouraged to collaborate and communicate with their peers, teachers, and adults outside the classroom. These activities are considered critical for students to develop teamwork skill, share ideas, and co-create knowledge and experience with their classmates. Community action projects were recently applied in teaching biotechnology [11] and nutrition [12]. Therefore, the report shows how PBL approach was employed for the design and implement of community project and discuss what the authors have learned. In this course, students were challenged to connect to the real- world practices through the community project. Through working in the action-based project, students will learn how to transfer food security ideas from class to the "outside" community. This work will serve as a meaningful foundation to support student's ownership of learning and their inquiry learning.

2. Research methodology

2.1 Overview of community project

According to Bloom's Taxonomy, the highest form of learning is applying knowledge in real-world challenges which can link the knowledge acquired in class to the surrounding community (Figure 1). Students follow framework given in each step to create an action plan at the end of the community project (Table 1). Students could work in groups of 3-5 people; however, students could work independently if they had instructions' permissions. In the data collection step, students were required to spend about 2-5 hours/project, either do research or site interviews. Classroom time consisted of two main components: 1) Lecture knowledge related to food security, 2) Instructor answered questions and gave feedback to students. The instructor also clearly communicate the purpose of implementing community project, that may help students feel engaged to the project (Table 2).

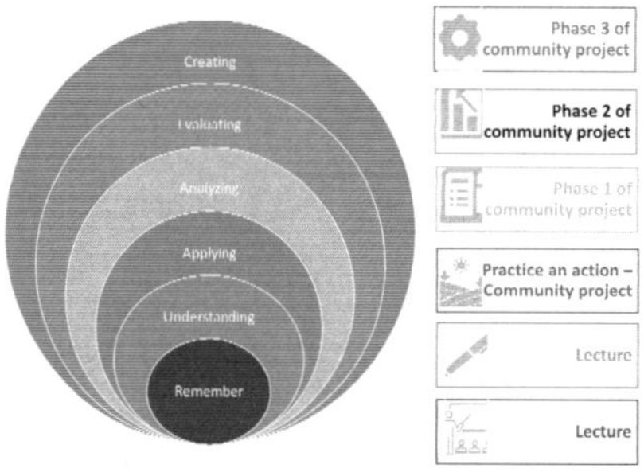

Figure 1. Each Bloom's objectivity linked to structure of the course

2.1.1 Phase 1 – Choose a topic and a community

Students chose a favorite topic related to food security and choose a community that might be related to the specific topic. To aid students, potential examples and a list of topics were provided to students for considering. The topics were shown as follows but not limited to:

- Genetic modified foods and their roles in solving food insecurity
- Food waste
- Food loss
- Water safety
- Food access for vegetarians
- Access to safe and nutritious food
- Access to food to maintain a healthy weight

In addition, students were provided some examples of possible community group that may be associated with the topics. The topics include (but not limited to) schools, local colleges, campus dining halls, markets, grocery stores, neighborhood, vegan community, restaurants.

2.1.2 Phase 2 – Learn the community and collect necessary data to access the needs of the community

In this step, students assessed the needs of the community that was selected from the previous step. Students can use different approaches – for examples, online platforms (email, google form, flyer printing, etc.), site interviews, or on-site visits. Although a major investigation was not necessary, students had to be familiar with making data-driven decisions.

2.1.3 Phase 3 – Action and final project

Next step, based on the data analyzed from step 2, students plan the action that will influence the community. The influence means change in opinions or habits of the community. The action may consist of providing information, education, or advocate some changes to the community group. In this step, students needed to demonstrate their action forms including power point slides, flyers, posters, video clips. In addition, students were also required to submit proof of action steps to the community, including photos, videos, email responses, etc. Finally, each project team submitted a report package, including includes a written report summarizing phase 1 and phase 2, demonstrations, and implementation of their action on the selected community.

Table 1. Framework of the Project

Phase 1- Topics and communities	- Select a topic and a specific community that student plan to invest to. - Plan to assess how knowledgeable the community is about the chosen topic. - Set up a platform to collect data from the community.
Phase 2 – Assess the needs of the selective community and plan an action	- Describe the interview process, surveys, questionnaires, and number of participants. - Attach all evidence such as video clips, forms, email responses, etc. - Present and interpret data, using charts and tables to present data, and explain how the findings help student planning the action. - Plan the action in which students describe what they will do to impact the community
Phase 3 – Action and final report	**Action** - Each group provides a copy of the group's action, which can be written such as power point slides, flyers, or posters, web page, etc. - Attach all document of the action, for examples, images, clips, emails, responses of community members. **Final report** - Write a final report of 10-15 A4 pages, 1.5 line spacing. - The final report should include data analysis from previous steps, community impact plan, and project results. - In addition, the report focuses on what students learned and student's recommendations for the project if they can continue the project in the future.

Table 2. The purpose of implementing community project

After finishing the community project, student should be able to assess the needs of a community. (i) evaluate the meaning of community service and the relationship between themselves and the community. (ii) demonstrate professionalism, integrity, responsibility, and courtesy in the classroom and in community engagement projects. (iii) practice communication skills in the context of community work and volunteering. (iv) make impacts on the community and evaluate the role of those impacts on the community.

3. Results and discussion

Adding community project in food security at an undergraduate level, this practice benefits students in multiple ways. Firstly, it encourages students to develop skills in 1) data acquisition and data analysis 2) conducting research questions and 3) working collaboratively. These are essential skills for the 21st century workforce [13, 14]. In addition, collaborative problem-solving skills are enhanced [13]. Numbers of student feedbacks showed their subject's interest and inspired ideas that connect the knowledge gained in the class to the "outside" world. Students believe that they can make an impact on community issues. Although students were not well familiar with the subject and teaching method, students have been successfully implemented after receiving detailed instructions and feedbacks from the educator. Below, two projects with short information about phase 1 and 2 were demonstrated. The final products were also attached to present their impacts on the community group (Figure 2 and Figure 3).

3.1 Example 1. Food Waste

Report of Phase 1
1. Topic: Food Waste
2. Community: College students.
3. Data collection plan and action description: The group assessed students' understanding of food waste in the world, estimated the daily amount of food waste, and student's behavior on wasting food.
4. Action plan: The group developed an information poster about the actual food waste in the world and provided advice and solutions to combat the waste of food. The forms of impact included giving information, communication, encouraging change in daily habits. The poster was presented (back and front) at the cafeteria or dining hall in university campus so student can read it when they enter the hall. In this step, students also submitted a contribution list of the team.

Report of Phase 2
1. Topic: Food waste in daily meals
2. Community of choice: Students studying at Industrial University of Ho Chi Minh City, Vietnam
3. Implementation time: 8 days.
4. Information collection: Participants submitted online survey questionnaires to the school's website (100 responses) and some of them had onsite interviews (100 responses).
5. Results: The project team found that around 40% of participant students sometimes waste food whereas a large portion of participants rarely waste food (51.5%). The students often bought excessive fresh foods and cannot finish until foods get unpleasant. Most of students do not fully learn how to reduce the food waste. 87.5% of participants chose to trash, 43.7% students stored the leftovers in fridge, and the rest chose to feed animals the leftovers. Only few proportions (11%) stated that they made a food compose at home.
6. Impact of the findings on their action plan: Because most of students understand the food waste issues, the project team focused on promoting waste-free actions, including introducing a food plan app (NUTRIONIX TRACK), proving handouts, and designing a media platform to convey the group's message about food waste. In addition, the group placed several leftover containers in campus dining halls to teach students how to turn food waste to organic fertilizer.

Report of Phase 3– Final products (Figure 2)
- Had a poster (front and back side) on the wall of the main dining hall.
- Handed posters (flyer version) to students at the campus and emailed the poster to participants who took the survey in phase 1. The goal was to provide student information about food waste and how to reduce the amount of food waste by changing daily habits.
- Created a Facebook page(https://www.facebook.com/Love-Food-Not-Waste-106521971289120) where the group could exchange ideas about food waste, kicked off "saving food" challenges, posted posters and short video clips, etc.
- Placed two food trash bins in the dining hall.

3.2. Example 2. Genetically modified foods

Report of Phase 1
1. Topic: Genetically modified foods (GMF) and their role in solving food security issues.
2. Community: College students.
3. Data collection plan and action description: The group assessed students' understanding of GMF including its definition, potentials in solving food hunger and food quality. The group also determined the need of foods in terms of quantity and quality at the student community.
4. Action plan: The group informed students of gene modified foods as alternatives, present benefits and concerns of GMF, demonstrated differences between native corns and genetic modifed corns, showed evidence of research, and provided scientific findings about GMF. The poster was presented at the campus as well as the group communicated with students during the presentation. In this step, students also submitted a contribution list of the team.

Figure 2. Artefacts of phase A) Poster B) Food trash bin placed in the dining hall C) Poster attached the wall of the dining hall D) Facebook page

Figure 3. Artefacts of phase 3: A) The front of the poster, B) The back of the poster, C) Flyer

Report of Phase 2

1. Topic: Genetic modified foods (GMF)
2. Community of choice: Students studying at Industrial University of Ho Chi Minh City, Vietnam
3. Implementation time: 8 days.
4. Information collection: The participants submitted online survey questionnaires to the school's website (160 responses).
5. Results: The project team found that students did not fully know about GMF as well not have enough information to allow them to make their choices on foods. However, most of them (83.1%) developed their interest in GMF and 76.3% student said they were willing to have GMF in their daily meals. There were 78.2 % student agreed that GMF is one of solutions for the current food hunger in their community as well as in the world.
6. Impact of the findings on their action plan: The project team focused on educating student about GMF and their benefits to combat food hunger and providing evidence about their current concerns. The group planned to collect GMF data and put together a poster to present to students.

Report of Phase 3– Final products (Fig. 3)

- Had a poster stand at the Institute of Biotechnology and Food Technology building and verbal communications directly with the community
- Handed posters (flyer version) to students at the campus. The goal was to help students have more information about GMF

Moving student participation to engagement, enhancing the student experience by using good quality ownership such as working in projects is needed [15, 16]. Students were encouraged to connect their leaned knowledge about food security to a community' needs through an action plan. Several activities within the project that seemed offer student opportunities to link their personal commitment and thoughts into food security research. For examples, students were provided ownership to selecting a topic in the first phase (Table 1). During the second phase, students constructed their own thoughts about findings and interpret them in positive ways toward the community. Finally, along with problem-solving skills and in depth- understanding of community issues, students can reflect their knowledge and gained experience in communicating (both verbal and written) and explaining scientific topics to public audiences through their presentations and multi-media products.

The community projects highlighted students' collaborations and team-working skills. Students developed concepts, ideas, and experiences and shared them to their peers while working on projects. Consequently, this offered opportunities for students to connect and may build relationships with their peers and community members.

Instructors need to ensure student commitment through intrinsic motivation that beyond in- class credits. The action step of the community project triggered their own interests and linked them to curiosity and creativity that are beneficial to the learning process. In this step, student designed and create products that can make a change in their group community. Students worked not only for their sake of their grade but also for a purpose of making a better community.

The PBL action project seems to have benefits to students and community, however, it is found that several limitations that may hinder student's inquiry learning. First, students may find difficulty in choosing a community to work with. They attempt to pick convenient groups that are their friends, neighbors, or clubs. This regard is not a serious limit but sometimes the selected community is not well-reflected on their chosen topic. Instructors may help to easy this challenge by having additional office hours to assist students and give them instruction in need. Data collection while assessing the needs of the selected community sometimes also hinder students. By the time of this course, the third wave COVID-19 outbreak hit Vietnam hardly, students found difficulty to directly contact or interview members in the community. This situation had a similar impact on students in designing an action plan. Therefore, adopting several online platforms should be helpful to aid students and the community as well. For examples, students employed multimedia tools to deliver their action plan, including Facebook page, emails, and webpage. Since, communities become more educated through accessing scientific information, a collaboration between the institute and some local community groups would be a critical step to observe further impact of this approach. The partnerships between the institute and those community groups may lessen burden for students on finding a particular group to work on each semester.

Regarding to instructor's effort, guiding students for the project plans, and grading different topics can be time consuming. In addition, fair grading individual over group performance would be challenging, therefore, a clear rubric, formative evaluation information from students about the project (such as phase 1, phase 2, and phase 3 of the project), and peer evaluation form are helpful for instructors.

4. Conclusions

Through implementation of the community projects, it is found that the incorporation of a semester length project into our food security course enhanced engagement among students. In this project, the student's ownership is highlighted, where they are involved in project design, and they make decisions for themselves. In addition, working with others outside the classroom, such as other faculties, students and people from the community, students develop teamwork skills and form connections with the real world. Based on this report, we recommend that providing students clear project guidelines, and using a variety of assessment including self and peer evaluation about the project would be helpful for the successful adoption of a PBL approach in food security at undergraduate level. More than 74% student graded "good" and "excellent" for overall quality of the course and 97% students recommend this course for others. Although this community project focused on food security, this approach can be applied in other fields.

References

[1] Capone, R, Bilali HE, Debs P, Cardone G. & Driouech N 2014. *Journal of Food Security*, **2** 13-22
[2] Cole MB 2020. *Teacher India*, **14** 1-3
[3] Raji PK, Sankar CS. 1999. *Journal of Engineering Education.* **8**. 501-508
[4] Anazifa R, Djukri D 2017. *Journal Pendidikan IPA Indonesia* **6** 346-355
[5] Hugerat M 2016 *Learning Environments Research* **19** 383-395
[6] Kokotsaki D, Menzies V, Wiggins A 2016 *Improving schools* **19** 267-277
[7] Kubiatko M, Caculova I 2011 *Energy Education Science and Technology Part B: Social and Educational Studies* **3** 65-74
[8] Wiek A 2014 *International Journal of Sustainability in Higher Education* **15** 431-449
[9] Lowire T, Leonard S, Fitzgerald R 2018 *Educational Design Research* **2**
[10] Marz RW 2004 *Journal of research in Science Teaching* **41**1063-80
[11] Nguyen PD, Siegel MA 2015 *The American Biology Teacher* **77** 241-247
[12] Zinger L, Sinclair A 2008 *Journal of College Teaching & Learning (TLC)* **5**
[13] Fiore SM, Graesser A, Greiff S 2018 *Nature Human Behaviour* **2** 367-9
[14] Suarta IM 2017 *Advances in Social Science, Education and Humanities Research* **102** 337-342
[15] Dounas-Frazer DR, Stanly JT, Lewandowski H 2017 *Physical Review Physics Education Research* **13** 020136
[16] Kaput K. 2018. Evidence for Student-entered Learning (Saint Paul, MN: Education Evolving)

IOP Conf. Series: Earth and Environmental Science 1057 (2022) 012003 doi:10.1088/1755-1315/1057/1/012003

A case study of thermal and humidity sensations in the smart building

J Dziedzic[1], **M Cedro**[2] **and P Wojciechowska**[2]

[1]Jan Kochanowski University of Kielce, Uniwersytecka 15, 25-406 Kielce, Poland
[2]Kielce University of Techology, al. Tysiaclecia P.P.7, 25-314 Kielce, Poland

E-mail: mcedro@tu.kielce.pl

Abstract. Modern smart buildings are equipped with renewable energy systems and control units to optimise and minimise energy consumption. Thus, they enable to reduce CO_2 emissions and help protect the environment. In the article thermal and humidity sensations of students in the smart, modern building have been presented and discussed. The analyses were done in three lecture rooms, where students filled in questionnaires about how they felt. Simultaneously, indoor air parameters were measured (temperature, relative humidity, carbon dioxide level and etc.). The study enables to assess the thermal and humidity in the smart, modern building, where people tend to spend large amounts of time. Thus, the valuable data can be obtained regarding indoor environment and human responses there.

1. Introduction

Modern smart buildings are typically equipped with renewable energy systems such as photovoltaic systems, heat pumps, solar collectors and etc. The centralised control units with BMS (building management system) constantly monitor the activities within the building and can optimise and, thus, minimise the consumption of energy for heating, air conditioning and lighting purposes. Moreover, they are usually very well insulated and need less energy for heating and cooling. They also recover thermal energy in ventilation systems using heat exchangers. Consequently, smart buildings help to reduce carbon dioxide emissions generated during energy production and, thus, have a positive impact on the environment.

Although smart (also called “intelligent”) buildings have been built for a few decades, they are still quite rare in Eastern European countries as well as other regions of the world. Thus, indoor comfort regarding temperature and humidity in not well studied in various climatic zones. While such buildings are expensive to build and maintain, they should offer high comfort levels, but it is not always true, as presented with the data in the next section of the paper.

In the paper by Merabtine et al. [1] thermal comfort and the quality of indoor air were considered experimentally. A level of dissatisfaction was detected in the building. Besides CO_2 concentration was high at some places and certain time and exceeded limiting values. Dębska [2] focused her study on over one hundred and sixty students who filled in the questionnaires in rooms where the temperature was 19.3 - 27.6 °C. The tests showed that about 80% of the students thought that the environment was comfortable or at least accepted it. The CO_2 level reached 1255 ppm, so it was high. The relative humidity was typically fine, but some students considered the air as too dry. The article by Krawczyk [3] analysed thermal comfort in one lecture theater in the smart building. The tests proved that over half of people expressed neutral thermal sensation votes. Still over thirty percent were pleasantly cool or warm. It shows that almost 20% were dissatisfied. Moreover, the author considered the impact of human Body Mass Index on thermal sensation votes as well as challenged the Fanger model of thermal comfort, claiming that it was inaccurate in smart buildings. Krawczyk and Krakowiak [4] performed tests in two rooms during summer. One room was located in a traditional building and the next in the smart building. They proved that the most crucial issue is how buildings are managed in order to provide thermal comfort. In this paper, also the Fanger model proved to fail determine accurately the thermal sensations of tested people. Majewski et al. [5] showed that large majority of students in the classrooms located in the analysed smart building were dissatisfied regarding their thermal sensations (it was especially reported by women). The experimental work was done on over a hundred people in three classrooms during summer. Białek and Koltuk [6] presented test results of thermal comfort analysis in the smart building, where ca. twenty percent of students proved to be not satisfied with their thermal environment and considered it to be too warm for them. However none of the people regarded the temperature as unpleasant (they generally thought of it as being acceptable). The authors also considered lighting in that smart building and proved to be generally fine.

Ensuring thermal comfort to all the people in a given room in a very difficult task. It is dependent on many factors: mainly indoor air parameters, but also the proper design of the building, its thermal insulation [7], heat transfer issues [8, 9], control systems [10] and etc. Consequently, the current study will be focused on thermal comfort tests in three separate rooms under the same weather, so that it would be possible to verify if the performance of the smart building regarding thermal comfort is at the acceptable level.

2. Experiment and thermal tests' results

The experimental procedure was conducted in the smart building called "Energis". It belongs to Kielce University of Technology (Poland). The building was completed in 2012. It utilizes renewable energy sources such as solar panels, photovoltaic systems, wind turbine and heat pumps. It has been equipped with BMS (building management system). This system runs and controls various building services. Figure 1 presents the picture of "Energis".

Figure 1. Smart building "Energis" on the campus at Kielce University of Technology.

The tests were done in such a way that students received questionnaires and answered questions about their thermal and humidity sensations. As they were filling in those anonymous questionnaires a microclimate meter collected data on the physical parameters in the considered rooms. Three lecture theatres on the first floor were selected for performing the tests in the study. The microclimate meter Testo 400, which was equipped with the probes of air and globe temperatures, relative humidity, CO_2 concentration as well as air velocity, was applied in the current experiment. Figure 2 presents the meter on the tripod.

Due to the fact that the study is done in such a way that the questionnaires are filled in at the time of recording the indoor air data in each of the considered lecture rooms, it is possible to determine how the physical parameters of indoor air influence the sensations of people in the rooms of the analysed smart building.

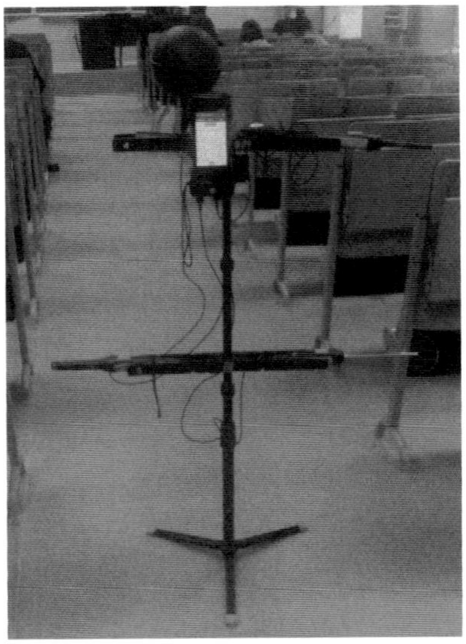

Figure 2. Microclimate meter in the lecture theater.

3. Results

The analyses have covered three lecture theaters located very close or next to each other on the first floor of the building. Their numbers are 1.09, 1.14 and 1.15. In each one the microclimate meter with adequate probes was used to determine the physical parameters of the indoor environment, but they were actually quite similar due to the fact that the air conditioning/ventilation and central heating systems were set at similar values in those rooms. However, there were some differences regarding carbon dioxide and humidity levels. These are quite dependent on the current and past presence of people in those rooms. The air temperature ranged from 21.6 °C to 22.2 °C, relative humidity from 30.8% to 34.7%, air velocity from 0.06 m/s to 0.08 m/s, while carbon dioxide level from 835 ppm to 990 ppm.

The students who voluntarily participated in the study expressed opinions about their individual thermal sensations (how they actually felt) using the anonymous questionnaires. The results for the three lecture theaters have been presented below in Figure 3. "Zero" describes neutral and, at the same time, the most favourable thermal state. The positive values (1, 2, 3) mean that the person feels warmth (1) or even considerable heat (3). On the other hand negative values (-1, -2, -3) mean the opposite – namely, cool and even cold sensations.

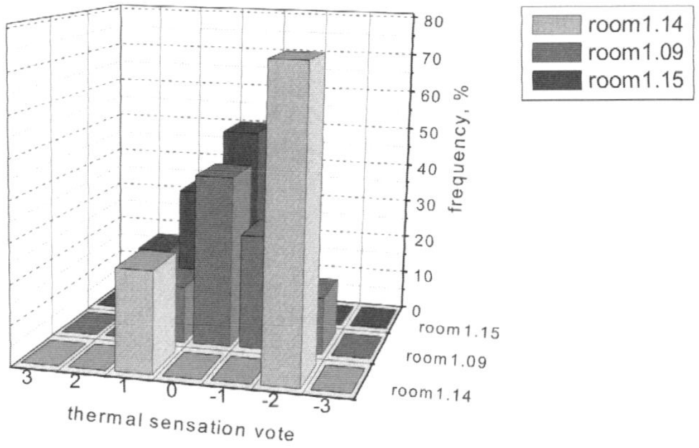

Figure 3. Sensations of students in three lecture rooms.

The air temperature in all the rooms was quite the same, but there were very considerable differences in the way how the students responded. In room number 1.09 over 40% of them felt fine (0 - neutral), but 14% reported that they were quite cold. However, in room number 1.15 about 16% of the respondents felt too warm. At the same time half of them considered the temperature as fine (0). Room number 1.14 was characterized by the biggest group of students who felt cold. At this location 3/4 of the people considered temperature as being low. Naturally, it can be caused by individual preferences of the students. Despite many negative voices regarding temperature none of those students wanted to change the temperature significantly as presented in Figure 4, which shows student preferences.

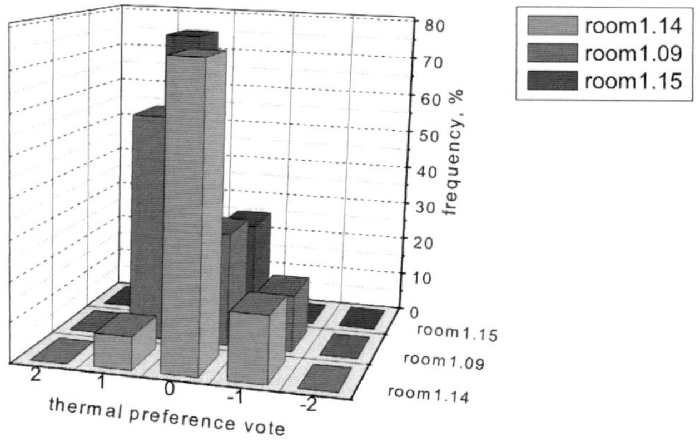

Figure 4. Preferences of students in three lecture rooms.

Students in room number 1.14 generally did not want any change (value of 0 in the figure) – 75% of them. Some of them would like the temperature to be higher (+1) or lower (-1). Different character of responses might be seen for rooms number 1.09 and 1.15, where the people would like to increase their air temperature (57% and 75%, respectively).

The humidity in the analysed rooms was quite low and ranged from 30 to 34%. Figure 5 presents the responses of students regarding humidity. The most favourable value is 0 (neutral state), while (-1) means that the air is too dry and (+1) that it is too wet.

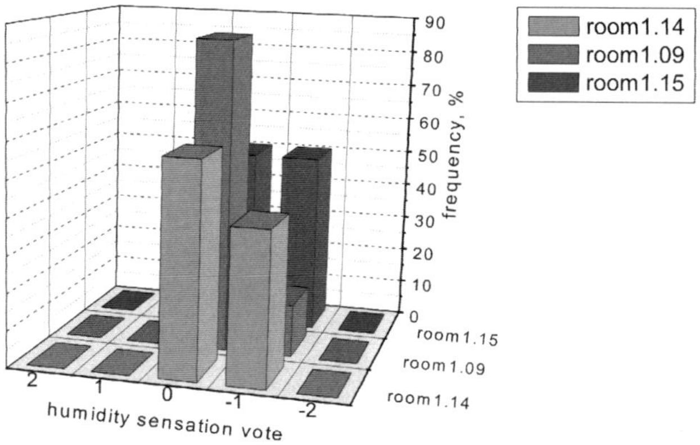

Figure 5. Humidity assessment in three lecture rooms.

The respondents assessed the humidity in the rooms as fine (86%, 58% and 50% of the answers in the questionnaires). The rest thought that the air was too dry. The values of relative humidity in the range of 30 – 34% can be considered as acceptable. Thus, the biggest number of the responses was positive. But for some people it is already too dry. This might happen especially when there are smoke or some other contaminants in the air.

Figure 6 below presents the individual preferences of the students related to the relative humidity in the considered rooms. If they were fine with this parameter, they answers 0 (neutral), but it they wanted to increase the humidity then the answer was (+1). The opposite would be (-1) for the preference of more dry air.

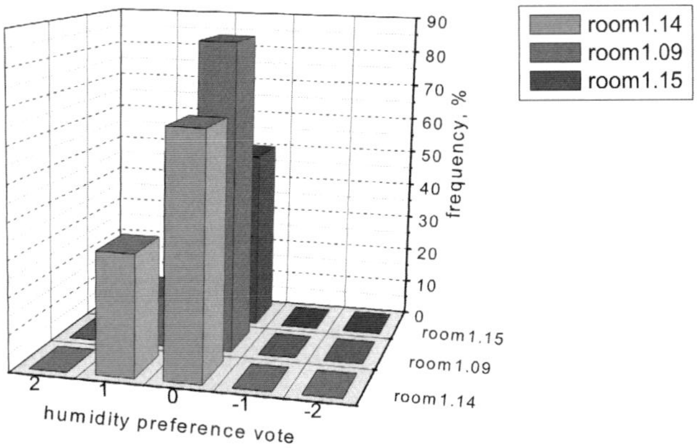

Figure 6. Humidity preference in three lecture rooms.

The figure is generally a reflection of the previous one. The students who thought that the air was dry wanted it to be wetter. The only exception is room number 1.14, where about 67% of students accepted the humidity and did not want any change, despite the fact that the votes of 0 (neutral) in Figure 5 were only 58%. However, the presented data show that humidity assessment differs from thermal assessment and is only influenced by the level of moisture in the air.

During the tests carbon dioxide level was determined. In order to check if it has any influence on thermal sensations of students, Figure 6 was created. It shows the relationship between CO_2 level and mean values of thermal sensations in each room.

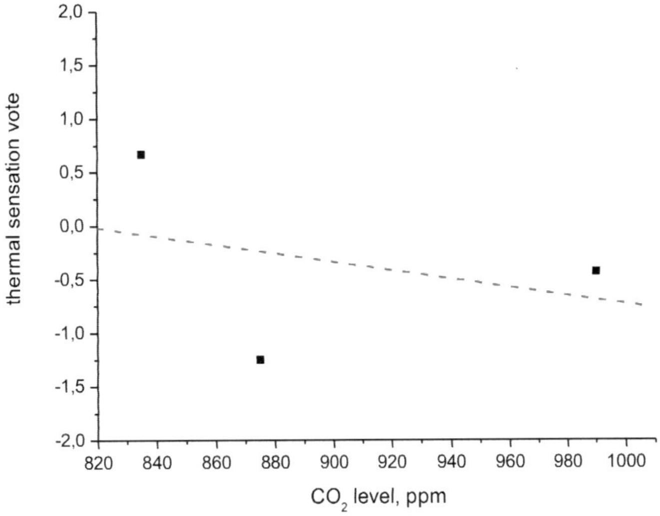

Figure 7. Relationship between CO_2 level and thermal sensations.

The number of data points is too low and should be increased in order to better determine the impact of carbon dioxide on thermal sensations, but it seems that there might be some relationship between those two parameters. This phenomenon requires further studies.

4. Conclusions

Smart buildings have a positive impact on the environment due to the reduction of carbon dioxide emissions. Thermal comfort tests performed in the smart building "Energis" provided additional knowledge on this important subject in this type of buildings. Generally all the students accepted the thermal environment in the three analysed lecture halls. They expressed small or no changes to it. But it was also observed that even in smart buildings the thermal and humidity conditions can be unsatisfactory for some people. It is clearly visible especially for humidity, because for many people it was too low.

5. References

[1] Merabtine A, Maalouf C, Waheed Hawila A A, Martaj N and Polidori G 2018 Building energy audit, thermal comfort, and IAQ assessment of a school building: A case study *Building and Environment* **145** 62–76 DOI: 10.1016/j.buildenv.2018.09.015

[2] Dębska L 2021 Assessment of the indoor environment in the intelligent building *Civil and Environmental Engineering* **17** 572 – 582 DOI: 10.2478/cee-2021-0058

[3] Krawczyk N 2021 Thermal comfort in the low energy building - validation and modification of the Fanger model *Proc. Int. Conf. Cold Climate HVAC & Energy (Estonia)* E3S Web of Conferences vol. 246. DOI: 10.1051/e3sconf/202124615003

[4] Krawczyk N and Krakowiak J 2021 The comparison of thermal comfort test results in selected traditional and modern buildings *Proc 10th Int. Conf. on Thermal Equipments, Renewable Energy and Rural Development* (Romania) E3S Web of Conferences vol. 286. DOI: 10.1051/e3sconf/202128602008

[5] Majewski G, Telejko M and Orman Ł J 2017 Preliminary results of thermal comfort analysis in selected buildings *Proc. Conf. on Interdisciplinatry Problems in Environmental Protection and Engineering EKO-DOK* (Boguszow-Gorce, Poland) E3S Web of Conferences vol. 17, 00056. DOI: 10.1051/e3sconf/20171700056

[6] Białek A and Koltuk B 2021 Thermal comfort measurements in the Energis building *Structure and Environment* **13** 10 – 15. DOI: 10.30540/sae-2021-002

[7] Piotrowski J Z, Orman Ł J, Lucas X, Zender – Świercz E, Telejko M and Koruba D 2014 Tests of thermal resistance of simulated walls with the reflective insulation *Proc. Int. Conf. Experimental Fluid Mechanics 2013* (Kutná Hora, Czech Republic) EPJ Web of Conferences vol. 67, 02095. DOI: 10.1051/epjconf/20146702095

[8] Dąbek L, Kapjor A and Orman Ł J 2016 Ethyl alcohol boiling heat transfer on multilayer meshed surfaces *Proc. 20th Int. Scientific Conference on The Application of Experimental and Numerical Methods in Fluid Mechanics and Energy* (Terchova, Slovakia) AIP Conference Proceedings vol. 1745, 020005. DOI: 10.1063/1.4953699

[9] Dąbek L, Kapjor A and Orman Ł J 2019 Distilled water and ethyl alcohol boiling heat selected meshed surfaces *Mechanics & Industry* **20** 701. https://doi.org/10.1051/meca/2019068

[10] Kolková Z, Hrabovský P, Florková Z and Lenhard R 2020 Analysis of ensuring thermal comfort using an intelligent control system *Proc. of XXII. Int. Conf. The Application of Experimental and Numerical Methods in Fluid Mechanics and Energy* (Slovakia) MATEC Web of Conferences vol. 328, 03017. DOI:10.1051/matecconf/202032803017

Acknowledgments

The scientific work in the paper has been possible due to the funding from Polish Ministry of Education and Science.

Antecedents and Consequences of Public Interest Litigations in Sustainable Environment Protection in India

Nikhil Yadav [1], Uma S Yadav[2], Anurag B Singh[3] and Geetu Yadav[4]
[1]USB-BBA, Chandigarh University, 140413, Mohali, India
[2]Motilal Nehru National Institute of Technology Allahabad, 211004, Prayagraj, India
[3] XIM University, 752050, Bhubaneswar, Odisha
[4]Regenesys Business School, 400013, Mumbai, India
[2]**Corresponding email addresses: usyadav@mnnit.ac.in**

Abstract. Environment protection is one of the important concerns for policymakers. In this paper, the author has investigated available literature and court decisions and found that Public Interest Litigation has been played a crucial role in Sustainable Environment Protection in India

Keywords: Public Interest Litigation, Environment Protection, Sustainable Environment,

1. Introduction

The environment has a broad term one term is defined as surroundings of physical arrangements of nature where livings and non-livings are present there. Environment "includes water, air, other living creatures, micro-organism, and the relationship between biological and no biological component which exists between air, land, water, and human beings, property, and plants". As per 1986 [17] Section 2(a), All-important resources used by humankind are gifts of nature so judicious use of these resources is an important concern for humanity because misuse or overutilization of resources can be as lead to critical environmental situations in our society [1]. To avoid the critical situation of the environment and proper regulation of environmental-related activities Indian governments have made several rules and regulations [2]. Same time society has avoided the age of nuclear power and cutthroat competition for economic growth so proper checks and balances of environmental laws further the role of the judiciary has become very important for public welfare [3]. Indian constitution article 39A provides provision for Public Interest Litigation (PIL) that empowered people to protect their interests and social justice [4]. Any person can seek justice in article 32 explained by Supreme Court under the article and in article number 226 in High Court. Supreme Court of India has given few relaxations in the provision of Public Interest Litigation (PIL). after this relaxation norm of PIL, it became an instrument of change in different matters like women's sexual harassment, prisoners' rights and corruption in public offices, and environmental protection further particularly environmental protection cases court has given several landmark decisions to improve environmental safeguard [4]. To understand the role of PIL we should try to discuss the advantages of PIL and who can file the PIL. This paper focused on the requirement for environmental protection laws and courts judicial remedies analysis that should be strongly available for environmental protection The Indian judiciary has made some remarkable doctrine. The hypothesis study will discuss more comprehensive and descriptive feelings about the environment further that will help in the policymaking and implementation by public officers [5]. This is not only limited to national importance but now it is a Global issue. It is truly all doubt is beyond this and without a neat and fresh environment, surviving life of people is not possible. This paper explains how the need for environmental laws to their working procedure by full of objects and meaning.

Justice Lodha has propounded that the judiciary is not for others, it is for people. If socioeconomic and ecological problems are increasing, then the Indian and even all other global courts have to play an active role to observe smooth and comprehensive environmental laws [7].

2. The situation of filing PIL and disadvantages of PIL

In the situation of avoidance and ignorance of environmental issues in India judicial activism of the Supreme Court has facilitated a green jurisprudence [8]. If there is a concern for the protection of the environment, the poor situation of heritage, and human beings, then PIL should be filed in any high court. We should file a PIL in case of protection of wildlife animals and crime with reserve diversity and biosafety. In critical situations, many times typical questions asked by the court from the society and the government further court has given so many landmark judgments to balance economic development and ecological status [9]. With the resolution of environmental issues has created some problem for economic development because so many unnecessary PILs has created hurdles in economic development but over the period Supreme court has taken so many corrective actions to discourage the unnecessary PILs. Over this period PIL becomes an instrument of good governance [6].

3. Review of Literature

When we observe different reviews of the literature then we found that major PIL approached in the case of wastewater of tanneries that were flowing in the river and cities of Tamil Nadu. that case which is famous by the name of Welfare Forum of Vellore citizen v. Union of India &others, ((1996) 5 SCC 647) [10], in this case, the apex court has ordered and declared the rule polluter pay in which untreated savage water of the industries and tanneries will be directly discharged after proper treatment in the Palar river. Polar River was the main source of water supply in Vellore city in Tamil Nadu. Another important matter is showing, Research Foundation RSTF for and Natural Resources Policy v. UOI, ((2007) 8 SCC 583) [11]., in 2005 in this case the petitioner has filed the PIL in SC that invoked the fundamental right of the citizens said in the article 21. When the French ship Clemnensueau posed a threat to the marine environment in the state of Gujarat, in this case in 2011 salon court put forward the order that polluter pay rule so the industry should follow the system but the appeal has dismissed. Other cement plants challenged the Himachal court order (Deva S. 2009)[5].__Sahu, G. (2008)[4] in his paper titled "Public Interest Environmental Litigations in India: Contributions and Complications" said how PIL has transformed the welfare and protection system of the environment. Further, in 2015, it highlighted the importance of the apex and reliable court in the preservation of the ecology and environment of India and provided some crucial solutions for the protection of the environment Sisodiya. A (2015)[1]. Srivastava.S. 2019 has explained emerging issues relating to Environmental changes in India and the World explained the different issues related to environmental changes at the global level and recommend some solutions.

4. Overview of Environmental PILs and their contribution to environmental Protection

In the Judiciary system, first-time environmental protection PIL was "The Rural Litigation & Entitlement Kendra (RLEK) vs Government of India" purpose of this PIL was to resolve the ecological imbalances created by lime-stone crushers units. Despite several economic losses court had given the order to close some units of limestone quarry in favour of environmental protection (Sahu, G. 2008 b)[4]. "Union Carbide Corporation v. Union of India" in the case of Bhopal gas tragedy case in 1983 was a drastic and tragic event in Indian history. In this case, the court had given the order for each individual for their loss because it is the absolute liability of the enterprise to protect human beings. In this case, first time SC has given the concept of absolute liability. Apart from this landmark judgment some other cases in which the Court had given innovation in the Judiciary to protect the environment like in "the given concept of polluters to pay concept. "M.C. Mehta v. Kamal Nath (1987)" [12] given the concept of the public trust doctrine, "Rural Litigation and Entitlement Kendra v. State of UP" Judiciary had given the theory of sustainable development, and "Narmada Bachao Andolan v. Union of India and Ors" the court had given thought of the right to a sustainable and healthy ecosystem. A green environment is a basic right for the basic survival of human beings **(Sisodiya 2015) [1]**. In light of the above different PILs courts had given several new innovative concepts of environmental protection and norms for paying compensation to victims. This is also realizing moral and legal obligations to all stakeholders of the society. Another side in the Indian judiciary was not able to enforce Directive Principles but different verdicts of courts in PILs have provided a strong instrument of laws to the public for indirect enforcement of Directive principles of state further PILs also provided laws against social injustice like poverty, misuse of power by governments and many more (Deva,2009).

4.1. Problems and Prospects: Period from 1970 to 1990 many PILs brought changes in the area of environmental protection and social justice but after the Liberalization of the economy in 1991 nature of the economy has changed from socialism to a mixed economy. These changes have given birth to many disputes between development entities and environmental protection. Same time the back of the curtains of environmental protection few PIL activists, and Non-Government Organizations (NGOs) started to participate in PILs cases for their personal and political interests like bargaining, or other favors. which given the negative impact on economic activities like blockage of the capital of investors, etc. Later on, it had corrected by the court to impose a penalty on fake PILs and admit only genuine cases but there is still some problem like the time taken the judicial process and state neutrality exist towards given verdicts. The environment should be on priority to our parliamentary system as well as the government because the directive principle of state policy in article 48-A says "The State shall endeavor to protect the ecology and the environment, to provide security and assured for the forest and wildlife of India." Further, our constitution also highlights the duty of citizen and responsibility through Article 51-A (g) which say "It will have to responsibility and accountability of every people of India to protect and secure forests, wildlife lakes, and to have compassion for the living being in-country". In the case of Sachidanand Pandey vs State of West Bengal, the Hon'ble Supreme Court of India said, "the Judiciary is accountable and responsible and bound to bear in mind Article 48 if any problem of environment & ecology is brought before the court as well as article 51(g)". Through the given verdict court has mentioned problem-related to the environment resolving the dispute is a priority of the court but balances and imbalances in environmental policy are a matter of state so in the policymaking process state should be also concerned with article 48 –A. As above mentioned courts' verdicts on PILs and constitutional provisions empowered the public to take safeguard the environment but also imposed accountability, and moral responsibility on the State and citizens as well. Often, environmental protection PILs become causes for blockage of projects, which leads to losses of economic resources. To avoid economic losses and biomass loss to save the wildlife of every flora and fauna and achieve economic efficacy, the State should avoid ambiguous laws and there should be a reflection on article 48 in the policymaking process. Regular inspection of those critical projects that were in dispute and the court had taken many corrective actions because many times court verdicts lacking behind in the implementation process by State.

5. Discussion

5.1 Recent case and effect of PIL on the protection of the environment

Due to the increase in population and climate change, awareness conflict increased so for the protection of flora fauna & wildlife. Several civil society NGOs have filed PIL. The PIL is increasing day by day some landmark case discussions are given below.

In this case, Bombay HC on 4th October 2019 has given the decision about the conservation of Aarey forest reserve and it became the famous case that is "Vanashakti and 4 others vs. Union of India" to conserve the Holy forest (forest of Aarey) located in Mumbai(Maharashtra) that forest was cutting for construction. It has been observed that for the environmental protection aspect, with the need for care and requirement of security for society. Many environmentalists opposed the government and the court denied the case content by hearing the fact of the circumstances. The same condition has seen during the case filed by "Bombay environment action group vs. Maharashtra state". As we know the river and ocean, the biodiversity of the mangroves creates a strong ecosystem of coastal areas of the ocean and hence best ocean habitat in 2019. In this case, due to deforestation of mangroves in provinces of Maharashtra, a PIL has filed for this committee constituted by the government in order of the Bombay high court. (BEAG that is *Bombay Environmental Action Group Vs Maharashtra state and others (17 September 2018)*. Odisha high court held that was NGT established for the protection and safety of environmental rules and regulations further make sure preventive action in case of violation of environmental laws. In this (BEAG) case, the national green tribunal explained the role of (NGT) in the matter of different cases environmental issues [18].

5.2 Polluter pay principle

This principle has been developed for solving the problem in the rapid industrial establishment and says that any industry that will pollute mire will pay more to fulfill the environmental loss and forestation about the balancing the environment. It has sometimes called to balance the natural law but industrial waste has a major factor to cause an infectious and serious challengeable problem to both components like living and abiotic of the environment so, to improve the biosafety and food products raised the living standard but posed adverse effects on the environment that have been occurred at the same time. In the case of the "Indian council of enviro-legal action vs union of India" principle of the polluter, pay has first time introduced in a legal environment. The court has decided the redemption of the polluter pay principle means the absolute responsibility of an entity or individual. if anyone harms the environment and ecology and it extends on living non-living purpose not only to a component of the victim of harmful pollution and cost of environmental restoring and component of the environment is the part of the complete process of sustainable development as such the same amount polluter is liable for paying the compensation [16].

5.3 Role of Environmental Impact Assessment

In the case of unnecessary PIL problems for environmental protection, the Supreme Court has decided that only those cases will be admitted for hearing that harm the environment that has been mentioned in our common future [19]. The further court stated offense will be decided based on the level of causality and hazards at the national or state level. If we observe another dimension of the judiciary,

SC has an important role in delivering justice that is not only for people's welfare but also especially for flora and fauna, wildlife sanctuaries, national parks, and biosphere reserves in India. In the prediction of the consequence of the new development that will help for new sustainable establishment in living biodiversity and even in society and it's a positive or negative influence on our environment. It tries to mitigate the problem that may occur further so that concerning public authorities may take strong positive and preventive measures. It ultimately returns the scope of PIL in the environment and is a beneficial method and reduction of conflict. It is the responsibility of every citizen living in that area and cares for, be compassionate toward living creators & wildlife animals, and conservation of biodiversity. According to the article, 51g of the Indian constitution under the PIL articles 32 and 226 of SC & high courts writ can be invoked

5.4. Care of environment by Indian constitution and interpretation of fundamental norms

If we see the article written related to environmental protection in the different constitutions of the world, then we found at the top republic constitution. India is one of the first countries in the world in which special laws are in India constitution in having special provisions for environmental conservation. Which is a basic right and a directive there are the principal and basic duties that inoculate the national committee to protect, improve and conserve the environment, and the court has interpreted and strengthened this mandate. Due to special provisions S, and C can give a wider interpretation of the environment in the country.

5.5. Every entity fully enjoys the provided right of the environment which is guaranteed by Article 21 (36) many government agencies apply environmental laws under the obligation (1996 2 J.T. (S.C.) 196.[13]

Polluter pay rule states that polluters will bear whole accountability if pollution spread is by it's it this is a basic principle that is part of environmental laws of the land requires remedial clean-up cost as well as the payable amount. Vellore Citizen's Welfare Forum Vs. Union of India, A.I.R. 1996 S.C. 2715, 2721. The protective principle reminds the government officers to participate and anticipate to prevent and attack the cause of environmental pollution this imposes proof on the development and industries in front of the environment.

5.5.1 Public litigation is one of the best tools to help the underprivileged people poor people exploited, and downtrodden millions of people. This is a collaboration and cooperative effort of the petition and the public authority to observe and take legal action in favor of vulnerable sections and communities and to provide social and environmental justice for their protection [21].

(a) Regarding PIL some initiatives of Recent developments in this regard. The current use of public interest legislation has become so famous and the number of PIL increased, as time has been spent and awareness about the significance of the environment and realized the number of PIL. Some other landmarks are also given (Decision was given by the supreme court on (4 October 2019) "Vanshakti and 4 other Vs and Union of India and 11 others". This order considers the needs of and requirements and needs of the society and the environmental aspect. NGT 2019 decision in case of CPCB to prepare the model for state and UTs for compilation of environment and that would help in the preparation of the country's environmental plan. This was directed by the NGT in the case of Astha Unmat and Shree Nath Sharma vs. Union of India and Ors. (2019 NGT)[15]

An Order had given in the case of Sailesh Singh vs. State of Haryana 2019. In this case, it was observed by the main bench of the National Green Tribunal that industrial development and the grave of human beings should not be at the cost of development. Another side is no compromise with water quality (case Saloni Singh Vs. UOI 2016). In the case of the Summit Mehta case and the UOI 2016 polluter pay principle that was invoked in that case, it was a matter of public importance and environmental justice. There is another principle that is no fault liability NGT has directed that a private entity is liable to pay compensation [14].

6. Conclusion:

As Mahatma Gandhi said, "need of every man is satisfied by Earth and it provides enough to satisfy, but it cannot satisfy every man's greed." Therefore, it is the responsibility of the State and citizens to make balances between full fill needs and desires. In this check and balances process, Public Interest Litigation has been used as a tool by citizens to protect the environment. As a democratic country Indian participation of the public in governance is one of the important antecedents of democracy and PILs cases, play a key role to empower the public in the democratic participation process. For a sustainable, efficient, and effective ecosystem, there are still so many milestones are remain to achieve like Public awareness about existing laws of environment protection and proper implementation of different verdicts of courts [20].

Acknowledgments

Authors wishing to acknowledge assistance or encouragement from colleagues, special work by technical staff, or financial support from organizations should do so in an unnumbered Acknowledgments section immediately following the last numbered section of the paper.

7. References

[1] **Sisodiya.A (2015).** The Role of Indian Judiciary in Protection of Environment in India. accessed 12/10/208 *Indian journal of law volume 49 issue(2)* https://www.lawctopus.com/academike/role-indian-judiciary-protection-environment-india/
[2] Ngosso, T. (2013). The right to development of developing countries: An argument against environmental protection?. *Public Reason*, *5*(2).
[3] Xie, L., & Xu, L. (2021). Environmental public interest litigation in China: A critical examination. Transnational Environmental Law, 10(3), 441-465.
[4] Sahu, G. (2008). Public Interest Environmental Litigations in India: Contributions and Complications. *The Indian Journal of Political Science*, 745-758.

[5] Deva, S. (2009). Public interest litigation in India: A critical review. *Sweet & Maxwell London access*12/10/2018https://pdfs.semanticscholar.org/8530/2cf2447c26d50a78a322b03f1af93a5e9fdd.pdf
[6] Deva, S., Public Interest Litigation in India(2009): A Critical Review Civil Justice Quarterly, *Vol. 28, pp. 19-40, 2009, Available at SSRN: https://ssrn.com/abstract=1424236*
[7] Kotzé, L. J. (2012). Arguing global environmental constitutionalism. Transnational Environmental Law, 1(1), 199-233.
[8] Gill, G. N. (2012). Human rights and the environment in India: Access through public interest litigation. *Environmental Law Review, 14(3), 200-218.*
[9] Gauri, Varun, Public Interest Litigation in India: Overreaching or Underachieving? (November 1, 2009). World Bank Policy Research Working Paper No. 5109, Available at SSRN: https://ssrn.com/abstract=1503803.
[10] Vellore Citizen's Welfare Forum Vs Union of India, A.I.R. 1996 S.C. 2715, 2721. the Indian Law Institute, Vol. 43: 2, (2001), p. 192
[11].Research Foundation for Science Technology and Natural Resources Policy v. UOI, ((2007)8 SCC 583), in the year 2005,
[12] M.C Mehta vs Union of India (1987) https://www.lawcian.com/post/m-c-mehta-vs-union-of-india-1987-air-1086A.I.R. 1987 S.C. 1086
[13] A.I.R(1996) 2 J.T. (S.C.) 196. cases on environmental law: supreme court of India *https://www.jstor.org/stable/i40220885 /https://indiankanoon.org/doc/1818014/*
[14] Abram Chayes,1 The Role of a judge in a Public Law Litigation in 89 Harv, L. Rev. 1281 (1976); Sheela Barsc Vs Union of India, A.l.R. 1988 S.C. 2211.
[15] Aastha Ummat July 1, 2020, *Emerging issues relating to Environmental changes in India and the Worldhttps://lawtimesjournal.in/emerging-issues-relating-to-environmental-changes-in-India-and-the-world/*
[16] Fertilizer Corporation Kamgar Union Vs Union of India, A.I.R. 1981 S.C. 344 at p. 355.
[17] Environment Protection Act,1986 is an Act of the Parliament of India. Accessed 12/10/208 http://envfor.nic.in/legis/env/env1.html
[18].Hima Kohli judges high court Delhi judgment 201014. Discussed, To sum up, the Indian judiciary has proved itself to be strategic and ordered to cement company of Uttarakhand.
[19] Our Common Future, (1987). pp. 220-221: Report of the World Commission on Environment and Development
[20] People's Union o democratic Rights Vs Union of Sofia, A.I.R. 1982 S.C. 1473 at p. 1477, Bandhua Mukti Morcha Vs Union of India, A.I.R. 1984 S.C. 802 at p. 811.
[21] Principle 15. David Freestone 'The Road from Rio: International Law After (he Earth Summit' in *Journal of Environmental Law. (1994), 193 at pp. 210-215.*

Secured Optimal Path to Identify the Networking Model using Cold Chain Logistics in Hospital Environment

G.Vijaiprabhu[1], P. Pandia Rajammal [2] , S. Rema Devi [3] and P.Karthika[4*]

[1]Assistant Professor, PG & Research Department Computer Science, Erode Arts and Science College (Autonomous), Erode -9. gvprabhu7@gmail.com

[2,3&4*]Assistant and Associate Professor, Department of English and Computer Applications, Kalasalingam Academy of Research and Education, Krishnankoil, Tamil Nadu, India. [4]*karthikasivamr@gmail.com

Corresponding Author: [4]*karthikasivamr@gmail.com

Abstract. Logistics refers to moving commodities from one location to another, typically from a manufacturing facility to a client. A logistics network is a collection of activities that include product creation, production, and distribution. Chain of custody Logistics are items that must be carried chilled from the manufacturing facility to the customer. Because of work with logistics networks, a secure networking model is required. This post will look at an intelligent secured networking model for determining the best route for cold chain logistics to a hospital setting. The best and shortest path between points A and B is found using the optimal pathfinder. It also takes traffic and transportation costs into account. Medicines and vaccines that must store at a specific temperature are part of the cold chain logistics for the hospital environment. As a result, path optimization is more important in cold chain logistics than in other types of logistics in the hospital environment. The Bee-Ant Optimization Algorithm (BAOA) is proposed in this study to perform intelligent transportation to a healthcare setting. The suggested approach is compared to the Ant Colony Optimization (ACO), Bee Colony Optimization (BCO), and Neural Network Model algorithms already in use. According to the data, the proposed algorithm has a 98.83% accuracy rate for delivering logistics to the hospital environment.

Keywords: Hospital Environment, Cold Chain Logistics, Secured Networking Model, Optimal path, Bee-Ant Optimization Algorithm (BAOA), Artificial Intelligence (AI).

1. Introduction

Cold Chain Logistics is a system or a procedure that allows temperature-sensitive items and products to be transported safely along the supply chain. The relationship between temperature and perishability must be evaluated and accommodated. Cold Chain Logistics is used in hospitals to preserve vaccines at specific temperatures along the supply chain [1]. The cold chain logistics business is currently at a relatively early growth stage. The primary goal of this paper [2] is to develop an intelligent algorithm for cold chain logistics distribution optimization based on secure networking. It can utilize Cloud computing to link the cold chain logistics database to external clients who use cold chain logistics. As a result, any database connection terminal can monitor and update data [3].

2. Review of the Research

The origins of virtual reality technology can be traced back to the 1950s, some decades ago. Virtual reality technology and equipment have matured in recent years, thanks to significant advancements in computer graphics processing, computational power, and other technologies. The commercialization of mobile intelligence, the growth of mobile internet, and the future development of mobile virtual reality technology will all mature as diverse skillsets and technology improve. First, hardware manufacture will gradually achieve industrial scale and growth. The resolution, square meter refresh rate, latency, computation, and rendering capabilities are important virtual reality hardware technology markers. Second, as hardware products have improved, the interoperability of software systems and hardware has also advanced rapidly in technology [4]. When temperature and relative humidity conditions during shipping and storage are insufficient, the quality of these products might quickly deteriorate. It can detect temperature differences in warehousing, handling, and transportation. The industry's principal concern is studying and analyzing temperature gradients inside rooms with refrigeration, containers, and trucks [5]. In addition, cold chain logistics is a subset of logistics that refers to a systematic plan for refrigerated commodities to be kept in a low-temperature environment throughout the whole production, transit, storage, consumption, and marketing process. It is to ensure that the goods' quality is not compromised in any way [6]. The proposed path-finding model addressed the lack of information about the immediate environment, the control-feasibility of the examined paths, such as the sharpest possible turns a vehicle can perform, and the computational demands of a time-dependent environment. [7]. A new diversion rule has been formed due to the cascading dips in the process under the edge failure situation, and hence the distribution price related to the transportation speed. The objective function of a cold chain logistics transmission using the path optimization method is to reduce distribution costs. [8]. Cloud computing's information technology (IT) resources can be considered services. Cloud computing has had a significant impact on technical progress due to the new IT installation. Cloud computing can supply essential software for corporate administration while lowering IT and hardware and software maintenance costs. Cloud computing allows small and medium businesses to access professional IT solutions with minimal IT investment. A cloud-based cold chain logistics system can be employed to connect the database between cold chain logistics and external customers. As a result, any database connection terminal can monitor and update data [9]. Here is a cloud-based cold chain logistics system that has been developed. Configuration management, business development, data updating and transmission, control strategy, data gathering, and data calculation are some of the functional aspects of this system. Can improve cooperation between cold chain logistics and their clients, resulting in co-control of product sales information, increased cold chain logistics speed, and better all parties' interests [10].

3. Motivation of the Study

The safe storage, logistics, and management of temperature-sensitive items from one point of manufacture to the point of consumption are known as cold chain logistics. Vehicles were not employed in this procedure but rather warehouses with cooling systems. Foods like processed food, medicine, blood, eye, and kidney are safely transported to a consumer with the help of cold chain logistics. This research report will look into the path optimization technique for cold chain logistics vehicles using an improved evolutionary algorithm. The logistics plan for the cold chain logistics vehicle is critical for reaching consumers in less time, distance, and expense. Data from road transportation is used to choose routes. For this, path optimization methods are applied. The Bee-Ant Optimization Algorithm (BAOA) is an artificial swarm intelligence technique based on the behavior of bees and bee colonies. The cold chain logistics path is optimized with the help of digital technology and the use of GPS technology. As a result, the optimum short way for transporting temperature-sensitive items is discovered. The primary goal of this study would be to reduce the amount of time, distance, and money spent on transportation.

4. Proposed Model

Figure 1 shows how Networking Models are established in advance to provide a smooth flow in the logistical movement that performs the best. The three most significant factors are discovered using networking models. They are, indeed.

- Customer satisfaction
- Functionality
- Cost Involved

Before building a logistic network, examine these variables to create an efficient and cost-effective logistics system. A variety of modeling strategies are utilized to provide the most excellent customer service while also making a profit and lowering logistics handling costs. This study uses optimization modeling to develop an intelligent secured networking model for determining the best path for cold chain logistics to the hospital environment.

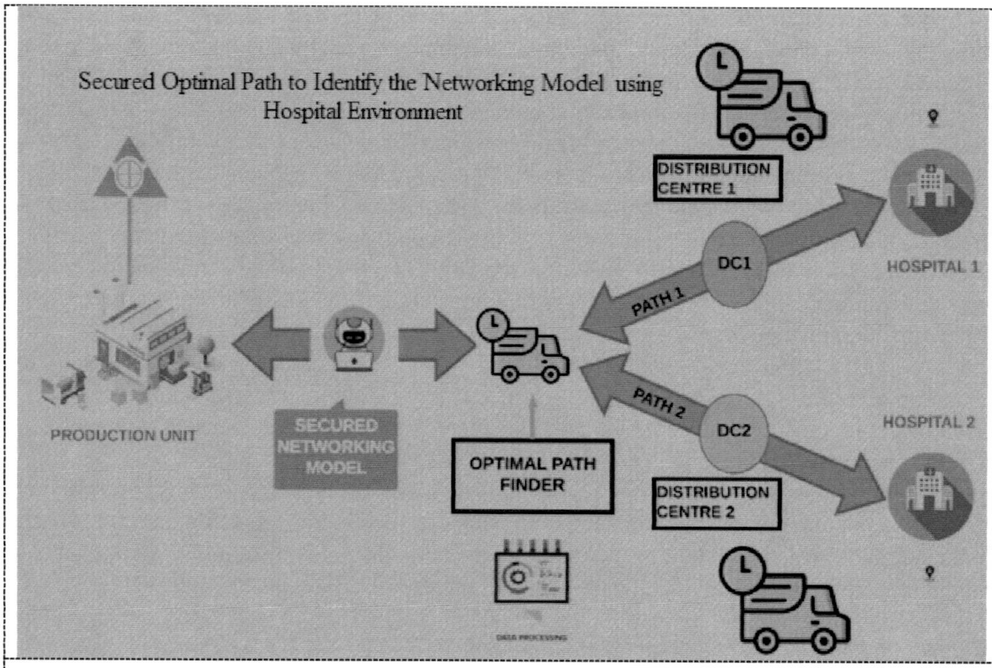

Figure 1. illustrates a proposed model for cold chain logistics in a hospital setting

The optimization approach employs mathematical formulas to find a better solution in the cold chain logistics to the hospital setting. This approach is based on assumptions such as the following:

- Number of production units
- Number of Distribution Centres
- Inventories must be maintained
- High customer satisfaction.

In logistics networks, linear programming is one of the most commonly utilized optimization models. This strategy entailed making the best use of resources to make the most money. The artificially intelligent computer systems are being used to implement the Linear programming method. This strategy is highly effective for matching supply and demand on both ends. As a result, items flow smoothly from manufacturing units to distribution centers and hospitals. As a result, path optimization is increasingly essential in cold chain logistics in the hospital setting.

Medicines and their associated equipment must maintain a specific room temperature to function correctly. Therefore, production units, distribution centers, inventories, and transportation systems should all be outfitted with the necessary standards in this respect when the product is in a static case, as in the first three places, it is easy to monitor the temperature. During transportation, however, the process is quite tricky. Therefore, the Cold Chain Logistics ideas are used to monitor the temperature while on the move and the Bee-Ant Optimized Algorithm to determine the best route for transit. The forward pass uses the Bee Optimized algorithm in the suggested approach, while the reverse pass uses the Ant Optimized algorithm. The optimal pathfinder must only focus on the path-finding process during the forward pass, which should be faster—created Bee Optimized Algorithm because Bee can discover the quickest way to the food supply from the hive's location. However, material transit is examined in terms of logistics during the backward pass. Chose Ant Optimization because the ant produces a chemical molecule called pheromone, which can evaporate. This concept can be linked to the cold chain property of temperature monitoring. If the temperature changes during transit, another logistics can be picked as an alternative solution to continue the procedure. As a result, the Bee-Ant Optimized Algorithm combination model is proposed.

5. Proposed Work

The optimization model appears to take a variety of influencing factors into account to accurately represent the actual situation, including time-varying transportation online traffic, traffic levels, customer time consumed, and the sleekness of innovative products. Still, it also considers the time it takes to charge a customer. The Bee-Ant Optimization Algorithm was devised to assist with Logistics Vehicle Electric Vehicle Routing (EVRP). According to simulation results, the proposed technique may effectively reduce delivery congestion, lower total distribution costs, and increase the efficiency of a cold chain logistic distribution system with new goods.

$\boldsymbol{K = P \cup M\{0\}}$ Denotes the set of all logistics nodes in the network. The letter $\boldsymbol{P}$ stands for "points of contact" with customers. The number 0 represents the processing facility; M represents a collection of rapid chargers, and K represents a collection of all endpoints in a logistics system. H represents the set of time intervals that occur throughout the day, $\boldsymbol{H = \{H_1, H_2, \ldots\ldots, H_n\}}$; $\boldsymbol{n}$ is the total number of time intervals. G signifies the number of road section kinds; G denotes the collection of road segments in the road infrastructure$\boldsymbol{G = \{G_1, G_2, \ldots\ldots, G_n\}; G.}$ $\boldsymbol{D = \{D_1, D_2, \ldots\ldots, D_n\}; D}$ is the number of different areas in the area set.
Let ρ_i denote the number of required charging stations capable of charging equipment utilization ratio $\rho_i = \left(\frac{\lambda_i}{x_i \mu_i}\right)$. The waiting period for such $g^{th}SK$ and attempting to capture point i is indicated by Equation according to the traditional slight Equation (1)

$$h_{ig}^{G} = \left(\sum_{n=0}^{x_i - 1} \frac{(x_i \rho_i)^n}{n!} + \frac{(x_i \rho_i)^{x_i}}{s_i!\,(1-\rho_i)}\right)^{-1} + \sum_{i=1}^{x} \frac{(x_i \rho_i)^{s_i} \rho_i}{\lambda_i x_i!\,(1-\rho_i)^2 \varphi_{D_i}} \times H_{ig} \times g^{th}SK \tag{1}$$

The load voltage for the $g^{th}SK$ at endpoint i in the charging procedure is as given in Equation (2).

$$h_{ig}^{G} = \sum_{i=1}^{x} \frac{(x_i \rho_i)^{x_i}}{s_i!\,(1-\rho_i)} + \sum_{i=1}^{g} \frac{S_{max} - S_{ig}^{H}}{n_h} \tag{2}$$

Not only does the vehicle's disposition affect its electric energy consumption, but so do its loads and speeds. The ability to run power $D(A_k, k)$ is represented by Equation (3) whenever an SK with a charge travels at A_k transportation distance ŋkm/h k on a flat route

$$D(A_k, k) = \sum_{i=1}^{g} H_{ig} \times g^{th}SK + \sum \frac{\left((y + A_k).\,g.\int k + (R_n.\,Z_i.\,k^3/22.56)\right)}{3600\text{ŋkm/h}k} \tag{3}$$

The actual speed of such a vehicle over a sufficiently short length of time to be proportionate to the determined duration of time. In addition, equation (4) has been revised to account for variations in the current roadways that the motorist intends to travel.

$$k_{ij}(t)=\sum\left((y+A_k).g.\int k+(R_n.Z_i.k^3/22.56)\right)\times\sum_{i=1}^{j}\begin{cases}k_{ij}^1, & t\in H_1,\\ k_{ij}^2, & t\in H_2,\\ k_{ij}^n, & t\in H_n,\end{cases}\quad (i,j)\in G \tag{4}$$

In x the allocation of speed fees on the street does not change in the same timeframe(i,j). A wide M range of factors are taken into account, including avenue kind, delivery period, as well as $M_1,M_2,\ldots..M_n$ time-varying velocity of the car, as well as such position and transition as described in the following Equation (5)

$$M=\sum g_{ij}(n_i-\bar{n})(n_j-\bar{n})\times\sum\{M_1,M_2,\ldots..M_n\}+\sum_{i=1}^{g}H_{ig}\times g^{th}SK \tag{5}$$

The transmitter and g_{ij} the receiver must be distinct from other vehicles, and the operator must be able to connect to the appropriate tool without difficulty (WSN with AI). The transition pace$n_i-\bar{n}$ is divided into on-the-spot transition and time change. The precedingMH, with the space-time transformation, takes a long time to conform with the series, which is represented by Equation (6)

$$MH=k_{ij}(t)\sum\{M_1,M_2,\ldots..M_n\}+\sum_{i=1}^{x}\frac{x\sum_{i=1}^{x}\sum_{j=1}^{x}+\sum g_{ij}(n_i-\bar{n})(n_j-\bar{n})}{\sum_{i=1}^{x}\sum_{j=1}^{x}+g_{ij}(n_i-\bar{n})^2} \tag{6}$$

The Cold Chain Logistics Vehicles based entirely solely during the transition time are represented by Equation (7).

$$MH=\sum\left(\int k+(R_n.Z_i.k^3/22.56)\right)+\sum\frac{x\sum_{i=1}^{x}\sum_{j\neq 1}^{x}+g_{ij}(n_i-\bar{n})(n_j-\bar{n})}{K^2\sum_{i=1}^{x}\times\sum_{j=1}^{x}g_{ij}} \tag{7}$$

Equation (8) then determines the number of storage media required to change the transaction based on the cold chain logistics time.

$$k_{ij}(t)=\sum_{n=0}^{x_i-1}\frac{(x_i\rho_i)^n}{n!}+\sum_{i=1}^{j}\frac{\sum_{j=1}^{q}\sum_{d=1}^{q}\sum_{u=1}^{q}\sum_{y=1}^{x_d}|s_{ij}-s_{dy}|}{2x^2t} \tag{8}$$

According to Equation (9), WSN network and AI technology work together to describe several phrases connected to diverse A vehicle kinds and data add-ons, and the amount of $s_{ji}-s_{dy}$ data may be represented by t_j+t_d by the focus arc vehicle speed and duration

$$petri_{jd}=\sum_{i=1}^{x\in s}\frac{(x_i\rho_i)^{x_i}}{s_i!\,(1-\rho_i)}\times\sum\frac{\sum_{A=1}^{d_j}\sum_{y=1}^{x_d}|s_{ji}-s_{dy}|}{x_jx_d(t_j+t_d)} \tag{9}$$

The equation (10) describes ϑ a WSN network with $\vartheta_n^2+\vartheta_x^2+B_1$AI-powered code operation processes B_1 Transformations like $2\vartheta_n\vartheta_x+B_1$ correlation between distinct τ objects of a particular statement and $2\tau_{nx}+B_2$ oriented action sequences have significantly diverse sorts of $\tau_n^2+\tau_x^2+B_2$ velocities and timings in the declaration.

$$M_{xn}=\sum_{i=1}^{g}\frac{S_{max}-S_{ig}^H}{n_h}-\sum\frac{(2\vartheta_n\vartheta_x+B_1)(2\tau_{nx}+B_2)}{(\vartheta_n^2+\vartheta_x^2+B_1)(\tau_n^2+\tau_x^2+B_2)}\quad t\in h_i \tag{10}$$

The modeling method of an influence of such code-based just on M_{xn} network on the optimization performance of the code behavior cold chain modeling based on the h_i network is also clarified by Equation (11); the v_q correlation between the y network component and the code component is also following Equation (12)

$$K(h_i,g_j)=\sum_{q=1}^{q}K(g_j/v_q)K(v_q/h_i)+\left(\sum K(h_i)K(g_j/h_i)\times\sum_{i=1}^{j}K(g_j/h_i)\right) \tag{11}$$

$$k_y = \sum_{x=1}^{x} \frac{2(x_2 - x_1)}{3} + \left(\sum_{x=1}^{x} \left[\frac{x_2 - x_1}{3} \right]^2 \times \sum_{y=1}^{y} \frac{2y}{y+1} + \left[\frac{1}{2} + \frac{1}{2y} \right] \right) \quad (12)$$

Static route analysis techniques are used to evaluate and process executable files to complete the evolution of $x_2 - x_1$ from rules to networks, as indicated in the following Equation (13)

$$P_{jn} = \sum_{x=1}^{x} \frac{2(x_2 - x_1)}{3} \times \int_0^{\infty} hS_j(x) \int_0^{s} (x-n) \times hS_n(x) \quad (13)$$

One programmer document $x - n$ associates a process M_{ih} with a specific input object set that has been processed, and the digital output object P_{jn} that is produced is reported as a result in Equation (14) for the process.

$$\ln\left(\frac{M_{ih}}{M_{ih} - 1}\right) = \sum_{i=0}^{h} \alpha + \beta \ln M_{ih} - 1 + (2\vartheta_n \vartheta_s + B_1)(2\tau_{ns} + B_2) \quad (14)$$

$$\sum_{n=1}^{\theta} Wn * l = \frac{\sum_{n=1}^{\vartheta} \left(\frac{\left(T + \frac{Y_i}{\sum_1^x D_\varepsilon} \right)}{M} + Y + YK \right)}{\beta + \alpha \ln M_{in} + \frac{K}{Y}} \quad (15)$$

Equation (15) represents the entire execution time, which is the sum of the unit time of $\sum_{n=1}^{\theta} \boldsymbol{Wn * l}$ and the total execution time. Time spent executing comprises $\boldsymbol{\beta + \alpha \ln M_{in} + \frac{K}{Y}}$ time spent mostly on highways and client points and time spent queuing and charging issues.

6. Results and Discussion

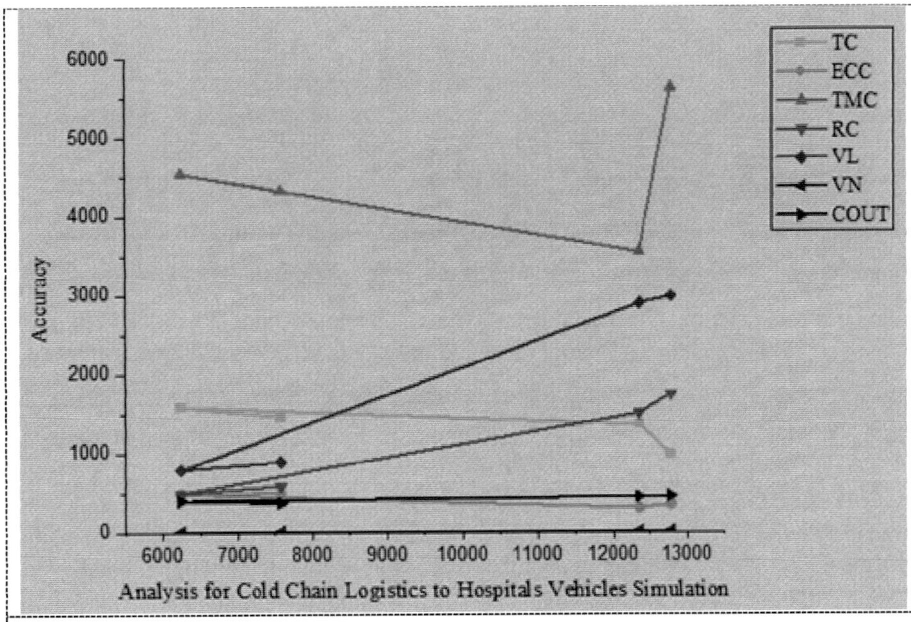

Figure 2. Performance Analysis Using Artificial Intelligence Techniques for Cold Chain Logistics to Hospital Vehicle Simulation

Let ρ_i denote the minimum requirements for charging stations capable of charging equipment with a capacity utilization ratio of $\rho_i = (\lambda_i / x_i \mu_i)$. There is a waiting period for these kinds of gth $g^{th}SK$. When attempting to charge a point i, it is described by Equation according to the traditional tiny Equation (1) shown in The performance study of the Cold Chain Logistics simulation employing WN and AI technologies for intelligent transportation is shown in Figure 2. Table 1 shows a numerical depiction of this figure in further detail. The following measures are used to assess the project: EX-, for example, is a Dataset Type. The abbreviation TC stands for Total Distribution Cost. Total Travel Distance (VD) and ECC (Energy Consumption and Charging), respectively, are abbreviations for Total Travel Distance and Energy Consumption and Charging. Time Management Cost is abbreviated as TMC. Refrigeration Cost is abbreviated as RC, Fresh Value Loss is abbreviated as VL, Vehicle Number is abbreviated as VN, and Running Time is abbreviated as CPUT (s). One of four dataset types is chosen at random in this investigation. In addition, the customer distribution data indicated before are used for evaluation.

Table 1. Analysis of Simulation Results in a Hospital Environment for Cold Chain Logistics Vehicles with Different Consumer Distributions

EX	TC	VD	ECC	TMC	RC	VL	VN	COUT
C103	13783.4	988.5	355.1	5723.6	1845.4	2995.54	9	454.3
C105	13355.2	1459.8	288.6	3647.3	1603.8	2992.93	9	443.9
R202	6445.67	1687.9	497.5	4634.2	498.2	799.32	8	395.4
RC203	7678.54	1567.3	484.3	4425.9	596.5	899.7	9	369.7

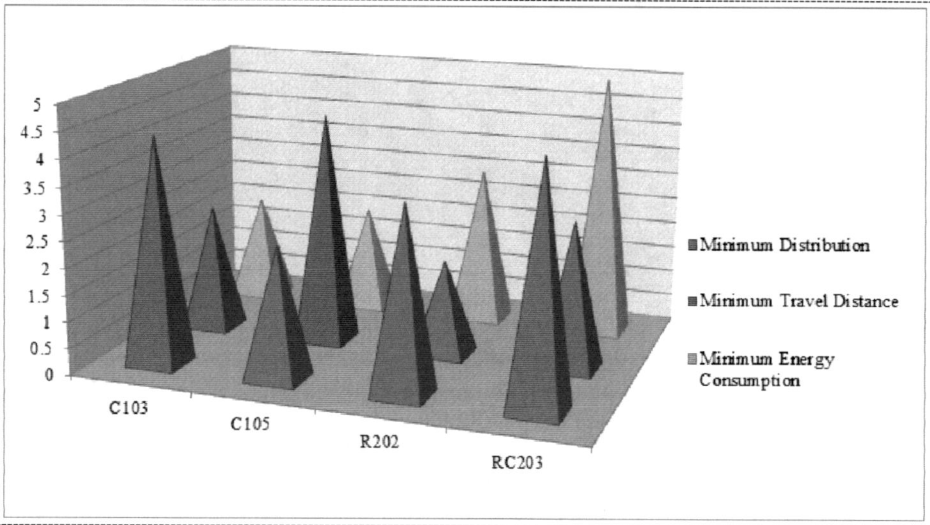

Figure 3. AI Techniques Used to Analyse Cold Chain Logistics Vehicles in Various Data Sets

The availability of speed fees on the road in x does not alter in the same timeframe as before (i,j). M variables are considered, including the street kind and delivery time frame, but also $M_1, M_2, \ldots.. M_n$. The constant-varying velocity of the car is connected with such much location and transformation as within the following. The data optimization study on cold chain logistics vehicles is depicted in Equation (5) in Figure 3. The optimization approach uses intelligent wireless sensor networks for online material ordering, which demands temperature control throughout resource transfer. It is accomplished by attaching smart sensor devices to the transport truck that continuously monitors the resources' status. The minimal distribution, journey distance, and energy consumption for transportation are shown in Table 2.

Table 2. WSN and AI Techniques Used to Analyse the Effectiveness of Cold Chain Logistics Hospital Environment Vehicles in Different Data Sets

	Minimum Distribution Cost (TC)			Minimum travel distance(Km) (VD)			Minimum Energy Consumption (kWh) (ECC)		
EX	TC	VD	ECC	TC	VD	ECC	TC	VD	ECC
C103	13783.4	988.5	355.1	14756.3	997.6	338.1	12234.9	990.2	356.7
R202	6345.67	1687.9	497.5	6334.2	1234.2	602.3	6567.3	1234.7	421.3
RC203	679.54	1468.3	4 95.3	669.3	36.7	78.1	565.1	667.2	99.1

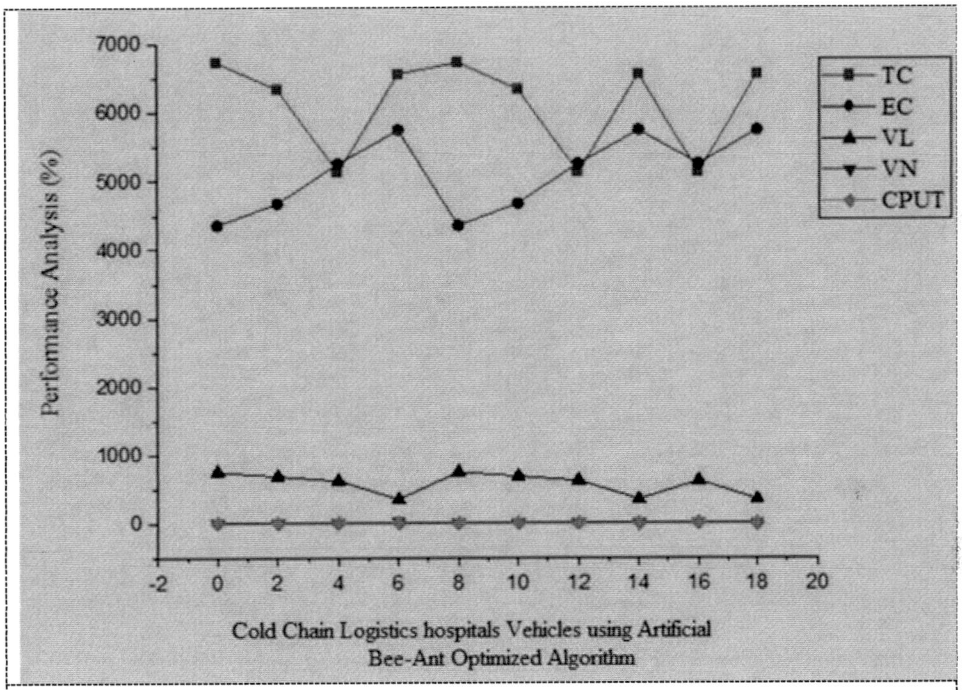

Figure 4. The Bee-Ant Optimized Algorithm analysed Cold Chain Logistics Healthcare Environment Vehicles in diverse data sets

A single program document $x - n$ associates a process M_{ih} with a specific filtered input item set and the digital signal object P_{jn} that is produced as a direct result of the process above Equation (14). The performance analysis depicted in Figure 4 is performed on various sizes. Table 3 displays the amount of data and the current state of the study. The corresponding analysis measures reach their peak values faster when the data size grows larger than when it is small.

Table 3. Vehicle Performance Analysis in Different Data Sets Using the Bee-Ant Optimized Algorithm in WSN with AI in Cold Chain Logistics Hospital Environment

Various data	TC	EC	VL	VN	CPUT
10	7734.1	5356.7	852.5	13	4

9	7345.9	5675.2	789.4	10	0
7	6136.1	6258.6	723.2	15	1
5	7567.9	6753.8	456.0	18	0

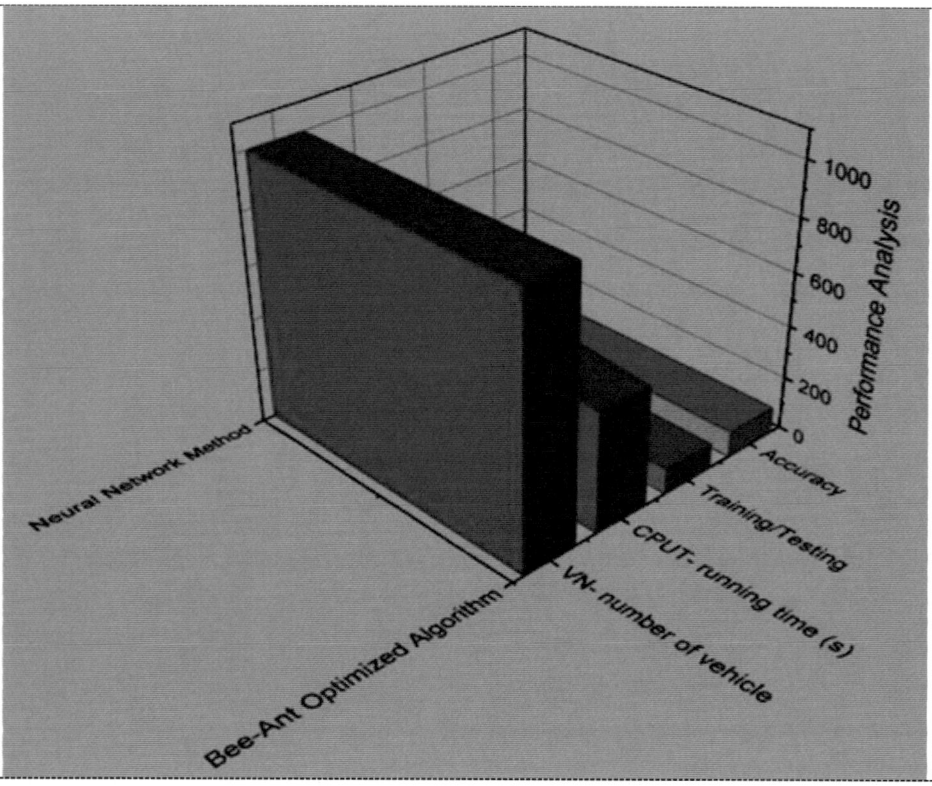

Figure 5. We assessed existing systems and Cold Chain Logistics healthcare Environment Vehicles in various data sets using the Bee-Ant Optimized Algorithm

The equation gives the maximum completion time and the quantity of the $\sum_{n=1}^{\theta} Wn * l$ unit of Time (15). $\beta + \alpha \ln M_{in} + \frac{K}{Y}$ Time spent mainly on roadways, but also at customer locations and effort expended not only in queuing but also at charging stations, is shown (Refer Figure 5). Temperature monitoring and various other comparable sensors are installed in cold chain logistics vehicles to ensure that they are constantly monitored. According to Table 4, 1000 hospital Environment vehicles are considered for evaluation in the Virtual Network (VN). In seconds, CPUT stands for the Time it takes to generate all feasible pathways. Following the artificial intelligence-assisted selection of the best path, the system will be able to train on that path while also testing the remaining transit modes. It has been revealed that implementing the ABC method leads to 98.83% accuracy in optimal path selection and cold logistics transportation along that path.

Table 4. We compared the outcomes of existing systems with Cold Chain Logistics hospital Environment Vehicles in different datasets using the Bee-Ant Optimized Algorithm

Algorithm	VN- number of vehicle	CPUT- running time (s)	Training/Testing (%)	Accuracy (%)
Bee-Ant Optimized Algorithm	1000	553.3	98.89	99.83
Existing Method: Ant Colony Optimization	1000	2427.9	1400.9	96.67
Existing Method: Bee Colony Optimization	1000	862.87	737.1	95.98
Existing Method: Neural Network Method	1000	663.6	94.23	94.54

From the point of manufacture to the end of consumption, cold chain logistics refers to the safe storage, logistics, and management of temperature-sensitive items. This technique did not include vehicles but rather warehouses with a cooling system. Cold chain logistics ensure that processed foods, medicine, blood, eyes, and kidneys are safely delivered to consumers. Researchers will investigate the path optimization technique for cold chain logistics vehicles using an updated evolutionary algorithm in this scientific article. The cold chain logistics vehicle strategy is essential for reaching customers less time, distance, and expense. Data from road transportation is used to develop routes. In various data sets, the Bee-Ant Optimized Algorithm is compared to the results of existing systems (95.54%) and Cold Chain Logistics hospital Environment Vehicles (98.83%). When compared to existing methodologies, our methods produce better results.

7. Conclusion

With the advancement of technology, an intelligent transportation system has become a requirement in today's world. An intelligent system's role must focus on all targeted users or clients. Furthermore, the things to be transported should be treated with care and routed in the most efficient way possible to ensure timely delivery. In recent years, logistics transportation has become more concentrated, as it contains a variety of medicines that must transport at a specific temperature. As a result, Cold Chain Logistics concepts are proposed in this research to monitor medicine transportation at a defined temperature and the Bee-Ant colony Optimization Algorithm to transport in the most efficient path. The findings show that the suggested model has achieved an accuracy performance of 98.83% in optimal path transportation. Compared to the previous Bee Colony Optimization, Ant Colony Optimization, and neural network model, this is higher.

Reference

1. Nedumaran Arappali, Melaku Tamene Mekonnen, Wondatir Teka Tefera, B Barani Sundaram, P Karthika, 2021, *Big Data Link Stability-Based Path Observation for Network Security*, Book: Computer Networks, Big Data and IoT ,p.459-468, Publisher: Springer, Singapore.
2. H. Liu, L. Pretorius, and D. Jiang, 2018, Optimization of cold chain logistics distribution network terminal, EURASIP Journal on Wireless Communications and Networking, p.no. 1, doi: 10.1186/s13638-018-1168-4.
3. Y. Chen, 2020, Intelligent algorithms for cold chain logistics distribution optimization based on big data cloud computing analysis, Journal of Cloud Computing, no. 1. doi: 10.1186/s13677-020-00174-x.
4. X. Li, Y. Wang, and X. Chen, 2011, Cold chain logistics system based on cloud computing Concurrency and Computation: Practice and Experience, no. 17, pp. 2138–2150, doi: 10.1002/cpe.1840.

5. L. J. Yao, J. Li, and J. J. Han, 2020, Application of virtual reality technology in intelligent cold chain logistics system, Journal of Physics: Conference Series, no. 1, p. 012030, doi: 10.1088/1742-6596/1651/1/012030.
6. L. Ruiz-Garcia and L. Lunadei, 2010, Monitoring Cold Chain Logistics by Means of RFID, in Sustainable Radio Frequency Identification Solutions, InTech,
7. X. Xu, 2011, Quality Control and Management of Plum in Cold Chain Logistics, Contemporary Logistics, pp. 28–31, doi: 10.5503/j.cl.2011.03.005.
8. M Prathap, Gashaw Bekele, Melkamu Tsegaye, P Karthika,2020, *Analysis for Time-Synchronized Channel Swapping in Wireless Sensor Network*, Book: Micro-Electronics and Telecommunication Engineering, p.265-273, Publisher: Springer, Singapore.
9. I. S. Dolinskaya, 2012 , Optimal path finding in direction, location, and time dependent environments, Naval Research Logistics (NRL), no. 5, pp. 325–339, doi: 10.1002/nav.21492.
10. Z. Fan, H. Ding, and W. Zhang, 2020,"Research on Cold Chain Logistics Path Optimization Considering Cascading Failure, IOP Conference Series: Materials Science and Engineering, no. 1, p. 012031, doi: 10.1088/1757-899x/787/1/012031.
11. X. Li, Y. Wang, and X. Chen, 2011,Cold chain logistics system based on cloud computing, Concurrency and Computation: Practice and Experience, no. 17, pp. 2138–2150, doi: 10.1002/cpe.1840.
12. X. Li, 2016, Development of Cold Chain Logistics in China, in Current Chinese Economic Report Series, Springer Singapore, pp. 263–288.
13. Asedo Shektofik Ahmed, Hussein Abdumalik Ibrahim, B Barani Sundaram, P Karthika, 2021, *Small Scale Targeted Face Detection using Deep Convolutional Neural Network,* Fifth International Conference on I-SMAC (IoT in Social, Mobile, Analytics and Cloud)(I-SMAC), P. 889-893, Publisher-EEE.
14. Venkatesh Andavara, Barani Sundaram, Didha Bacha, Tolossa Dadi, P Karthika,2021, *The Impact of Perceived Ease of Use on Intention to Use Mobile Payment Services for Data Security Applications*, Second International Conference on Electronics and Sustainable Communication Systems (ICESC), p. 1875-1880, Publisher-IEEE.

IOP Conf. Series: Earth and Environmental Science 1057 (2022) 012006 doi:10.1088/1755-1315/1057/1/012006

A Firebug Optimal Cluster based Data Aggregation for Healthcare Application

Sree Ranjani N Y[1], A.G Ananth[2], L Sudershan Reddy[3]
[1]Department of Electronics and Communication Engineering, JAIN (Deemed-to-be University), Bangalore, India.
[2]Department of Electronics, NMAM Institute of Technology, Mangalore, India
[3]Department of Management Studies, JAIN (Deemed-to-be University), Bangalore, India

Abstract. Majority of Wireless Sensor Networks (WSNs) are made up of huge sensor nodes with minimum cost, processing, communication capability and limited sensing. Because sensor nodes have limited resources, it's critical to enhance bandwidth utilization, network lifetime and minimize the data transmission. The term data aggregation defines the method of integrating the sensor data to minimize the data transmission. However, WSN-based healthcare applications face privacy, and security concerns. In this paper design of a Fire Bug Swarm Optimal Cluster-Based Data Aggregation (FSOCDA) in WSN has been presented which is a useful method for WSN to save energy and extend network lifetime. The efficient clustering is carried out using FSO, which is utilized to choose the CH from among various nodes in the cluster that serve as data aggregation nodes. By altering the network scale, the computational competence of the FSOCDA technique is compared to that of the non-aggregate approach. Further, the data aggregation method is helpful in healthcare applications in terms of data transferring from host hospital to other.

Keywords: Cluster Head, Healthcare Application, Wireless Sensor Network, Data Aggregation, Fire bug swarm optimization.

1. Introduction

Wireless Sensor Network is a self-contained system containing minor sensor hubs that can communicate with one another via radio signals, and the said sensor nodes may monitor, detect and comprehend the physical environment [1]. Sensor networks are generally used in hostile situations to gather and analyze low-level data in real-time. As a result, they're suitable for a wide range of surveillance applications and monitoring the healthcare applications [2]. WSN is continuously deployed in an untrusted and insecure environment, exposing it to all types of attacks and causing major security issues [3]. Furthermore, a high volume of information transmissions leads to congestion and data collisions. All of this has resulted in the development of data aggregation techniques. [4-6]. There are primarily three types of data aggregation methods.

The data aggregation first type clustering, this involves gathering and aggregating data at a cluster node before sending it to a sink node. The second approach is hop-by-hop aggregation, which collects data at each intermediary node. The partial aggregation is the third form of aggregation which happens when data aggregation hits a time or energy threshold. [7]. A cluster head always uses more energy than others in clustering data aggregation; WSNs usually employ a cluster head (CH) choose algorithm to choose a new cluster head, which trashes a lot of energy and time [8].

2. Literature Review

Saleh N et al. proposed the energy-effective dynamic clustering algorithm (EE-DCA) for the wireless sensor networks that enhances the bandwidth utilization via altering the action of the wireless scheme of the sensor node, remaining energy computation as well as the transition of state of the sensor nodes [9-12]. The dynamic clustering technique was utilized in enhancing the network lifetime as the clustering head's residual energy was lower than the threshold. The performance of this method was evaluated using the measures such as energy consumption as well as the throughput. The proposed method enhanced the technology for the early recognition and diagnosis of the irregular circumstances of the patient's health. The proposed method was effective and protected for the healthcare documents [13].

Rezaeibagha F et al. presented the effectively secure model to protect the data examination to handle the data examination for IoT sensor. The new cryptographic accumulator depends on the authenticated additive homomorphic encryption that collects the data from the IoT wireless devices. The aggregated data information regarding the patient's medical position was valuable for the doctors as well as the healthcare providers. The performance of this method was computed using the measures such as overall computation cost and computation time [14].

3. Health monitoring process

Initially, the patient's data was applied to the data aggregator using sensors such as blood pressure, pulse oximetry, ECG, and mobility, which are represented as sensing nodes. The data aggregator then aggregates all of the health information for data analysis[15-16]. The data aggregates every user health data as well as reports it to honestly provide medical server report; but it is concerned regarding individual sensor reading. The system model is shown in figure 1.

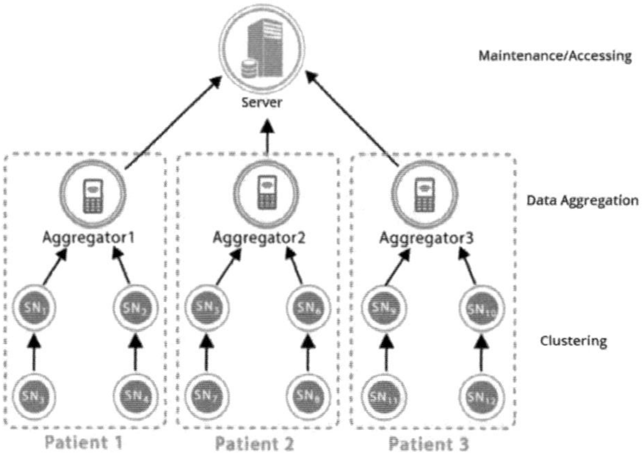

Fig. 1 Data Aggregation System Model

3.1 Authentication and Data Integrity:

Malicious operations must be detected in order to verify encrypted health information that are gathered and transmitted by genuine medical sensors and are not altered while transmitting data. Our proposed scheme's major security goal is to keep the data transferred by medical sensor to the private remote servers. The objective involves in preserving the data's integrity that are accomplished by data source verification. Furthermore, the suggested system should protect the anonymity and privacy of medical data without enabling any adversary to have access to it.

3.2 *Cluster development stage*

The entire system network is isolated into C layers; where every single layer is subdivided into cluster set. The cluster development phase is initiated by the base station [17-18]. Let us consider

NR and ω be the network region and the nodal distance that are located apart from the sink nodes. Then,

$$\omega = \max\left(\nabla_{j=1}^{M} e\left(se_j, BS\right)\right) \tag{1}$$

Where $e\left(BS, se_j\right)$ signifies the distance among the sensor node and the sink node se_i. Here, the sink node utilized two different layers for providing proper communication among the cluster head. In addition to this, C relies significantly on ω and the radio transmission containing minimum range $TRANS_{r-\min}$ is discussed in the following equation.

$$C = \frac{\omega}{TRANS_{r-\min}} \tag{2}$$

Let $l = \left\lfloor \frac{e\left(se_j, BS\right)}{TRANS_{r-\min}} \right\rfloor$; $se_j \in l^{th} layer$; where $1 \geq l \leq C$ (3)

The nodes containing minimum distance are referred as closer layer. The initial and the second layer are denoted as one hop and two hop distance; where in *l th* layer, *l* hop distance is present. It is noted that the hop distances are computed in accordance with the radio transmission containing minimum range. Finally, the M nodes fit in any of the C layer and hence every single layer is classified into K cluster. The layer containing minimum number of layers must contain high K value and the layer containing maximum number of layers must contain minimum K value. The total number of K clusters with respect to each layer l is stated in equation (4):

$$Kl = \frac{n(l)}{(l \times C)} \tag{4}$$

Where $n(l)$ represents the total nodes in the *lth* layer. In each layer l, Kl represents the total number of clusters employing K-means algorithm. A specific number of flat and disjoint clusters are generated using K-means clustering approach. Finally, *Kl* clusters are formed in this phase for all layers. The formation of clusters are executed to obtain optimal CH. This further minimizes the transmission of messages and enhances the network lifetime. An optimal cluster-based data aggregation selects an ideal node as a cluster head soon after getting the optimal clusters. The behaviour of Fire bug swarm optimization performs effective clustering.

For the security purpose, we proposed every CH have to perform safe aggregated data, hence the energy requirements were considered for performing the Bayesian fusion algorithm by describing the cost of communication.

CE_c = Energy to transmit a bits of data to its next hop cluster head + Energy to receive a bits of data from its cluster members + Energy to compute the trust probability for every cluster member

$$CE_c = E_{ty}\left(a_{se,CH_{mh}}\right) + \sum_{i=1}^{n(L)} \left[RE_{ry}(a) + \varepsilon\right] \tag{5}$$

Where, ε denotes the trust probability computation. If a node is selected as CH, it transmits the CH information to BS, cluster members and other CHs. Hence, every cluster members transmit their messages to CH on specified time period. The total trust is calculated by cluster head nodes thereby avoiding the calculation of power utilization on low energy cluster members. If the cluster head identifies unreliability of any node within the cluster, it alerts the left over cluster regarding malicious nodes. Hence the data from malicious node is discarded. In this concept, the malicious code is identified in small crowd and security threat is determined by power node.

Optimal cluster-based data aggregation method evenly distributes the load for the entire cluster nodes. The energy is dissipated by the nodes for static cluster head. If the alive nodes in a cluster are not greater than the obtained percentage then the cluster member join with other neighboring clusters. Also, we can minimize the overall computational cost and it is emphasized that by adding small cost, an effective end to end secure data aggregation technique is developed in heath monitoring systems.

4. Result and Discussion

We evaluate current methodologies to discuss the proposed FSOCDA technique's performance. In addition, the computational complexity is assessed in terms of latency, energy-aware data transmission, and cost in this section. The results of the FSOCDA (Firebug Optimal Clustering based Data aggregation) approach are compared to SPPDA (Secure privacy preserving Data aggregation) techniques by reducing communication complexity, computational overhead, and improving the effectiveness of the proposed method, which is implemented in NS2.

Table 1: Comparison of Computational Time vs Sensor Node

Nodes	FSOCDA	SPPDA
1	3	6
2	9	15
3	14	19
4	20	26
5	24	29
6	30	39
7	36	45
8	42	50
9	48	57
10	53	69

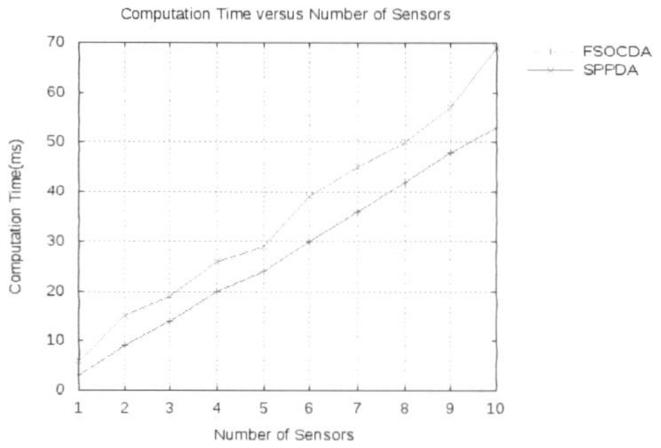

Figure 2 Computational Time vs Sensor Nodes

At the Server with FSOCDA and Existing schemes, the Computational time and sensor node comparison described in figure 2. At the aggregator and server, the computational time is highly minimized by proposing FSOCDA scheme as shown. The computational time and simulation results are more effective by using proposed FSOCDA model from the above analysis. The table 1 shows the computational time comparison at the sever for the proposed FSOCDA and existing algorithms.

Table 2: Comparison of Alive nodes vs Iterations

Iterations	FSOCDA	SPPDA
0	100	99
10	87	80
20	81	74
30	73	69
40	58	54

50	52	48
60	37	30

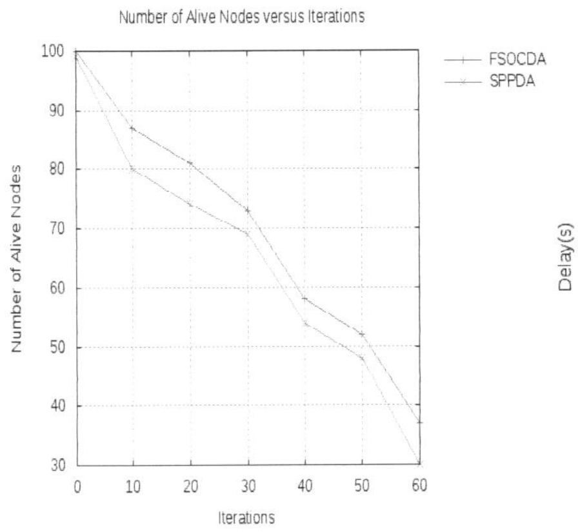

Fig. 3 Alive nodes vs Iterations

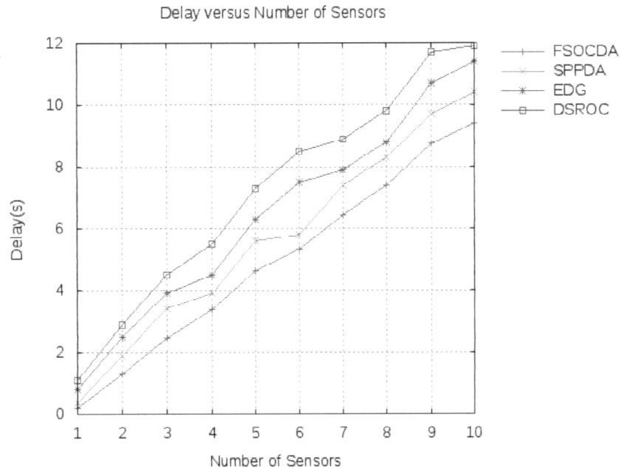

Fig. 4 Comparison of Delay vs Sensor nodes

By altering the network scale, the computing competency of the FSOCDA technique is compared to that of non-aggregate approaches. As a result, it is obvious that the FSOCDA technique outperforms the non-aggregate model. It is also obvious that for health monitoring applications, the best clustering-based data aggregation is utilised. Table 2 shows the comparison of Alive nodes versus Iterations for planned FSOCDA and current SPPDA. Figure 3 depicts the comparison of Alive nodes vs. Aggregator Iterations for FSOCDA and SPPDA.

The figure 4 shows the comparison of Delay vs Sensor nodes on Aggregator for FSOCDA and SPPDA. The comparison of Delay vs Sensor nodes for proposed FSOCDA and existing SPPDA is described in table 3. The results show better improvement comparing to the existing methodologies and proved to be the best.

Table 3: Comparison of Delay vs Sensor nodes

Nodes	FSOCDA	SPPDA	EDG	DSROC
1	0.2	0.35	0.8	1.1
2	1.299540408	1.9	2.5	2.9
3	2.467558123	3.4	3.9	4.5
4	3.373376488	3.9	4.5	5.5
5	4.652858361	5.6	6.3	7.3
6	5.335706006	5.8	7.5	8.5
7	6.416012521	7.4	7.9	8.9
8	7.388483941	8.3	8.8	9.8
9	8.750697558	9.7	10.7	11.7
10	9.421783011	10.4	11.4	11.9

5. Conclusions

1. The experimental results of FSOCDA method show better efficiency of proposed techniques.
2. The computational competence of the FSOCDA approach shows that network scale that it is 53ms compared to that of SPPDA has 69ms for 10 nodes. The FSOCDA approach is found to be better than the non-aggregate model.
3. The delay can be much reduced to 9.4s in the existing model when compared to the SPPDA.
4. It is concluded that FSOCDA model is much better than the SPPDA .
5. The proposed FSOCDA technique can be suitable for health monitoring application.

6. References

[1] Gajender, A., and M. Nagaraju. 2016 *Privacy-Preserving Data Transmission Protocol for Wireless Medical Sensor Data* IJITR 4, no. 6, pp. 4518-4522.

[2] Zhang, Changlun, Chao Li, and Yi Zhao. 2015 *A balance privacy-preserving data aggregation model in wireless sensor networks* International Journal of Distributed Sensor Networks 11, no. 6 pp. 937280.

[3] N. Nguyen, B. Liu, V. Pham and Y. Luo, 2016 *On maximizing the lifetime for data aggregation in wireless sensor networks using virtual data aggregation trees* Computer Networks, vol. 105, pp. 99-110.

[4] A. Muthu Krishnan and P. Ganesh Kumar, 2015 *An Effective Clustering Approach with Data Aggregation Using Multiple Mobile Sinks for Heterogeneous WSN*, Wireless Pers Commun.

[5] Hassanalieragh, et. al., 2015 *Health monitoring and management using Internet-of-Things (IoT) sensing with cloud-based processing: Opportunities and challenges* In 2015 IEEE International Conference on Services Computing, pp. 285-292.

[6] K. Parmar and D. Jinwala, 2016 *Concealed data aggregation in wireless sensor networks: A comprehensive survey*, Computer Networks, vol. 103, pp. 207-227.

[7] L. Villas, A. Boukerche, H. de Oliveira, R. de Araujo and A. Loureiro, 2014 *A spatial correlation aware algorithm to perform efficient data collection in wireless sensor networks* Ad Hoc Networks, vol. 12, pp. 69-85.

[8] Saleh N, Kassem A, Haidar AM. 2018 *Energy-efficient architecture for wireless sensor networks in healthcare applications*. IEEE Access. pp. 6:6478-86.

[9] Rezaeibagha F, Mu Y, Huang K, Chen L. 2020 *Secure and efficient data aggregation for IoT monitoring systems*. IEEE Internet of Things Journal, pp.8056-63.

[10] Al-Zubaidie M, Zhang Z, Zhang J. Reisch: 2020 *Incorporating lightweight and reliable algorithms into healthcare applications of WSNS*. Applied Sciences. pp. 10(6).

[11] Almalki FA, Soufiene BO. 2021 *EPPDA: An efficient and privacy-preserving data aggregation scheme with authentication and authorization for IoT-based healthcare applications*. Wireless Communications and Mobile Computing.

[12] Ullah A, Said G, Sher M, Ning H. 2020 *Fog-assisted secure healthcare data aggregation scheme in IoT-enabled WSN*. Peer-to-Peer Networking and Applications. pp. 13(1):163-74.

[13] Raja KN, Beno MM. 2017 *Secure data aggregation in wireless sensor network-Fujisaki Okamoto (FO) authentication scheme against sybil attack*. Journal of medical systems. 41(7), pp. 1-6

[14] Patan R, Ghantasala GP, Sekaran R, Gupta D, Ramachandran M. 2020 *Smart healthcare and quality of service in IoT using grey filter convolutional based cipher physical system*. Sustainable Cities and Society. pp. 102141

[15] Amudha S, Murali M. 2020 *Deep learning-based energy efficient novel scheduling algorithms for body-fog-cloud in smart hospital*. Journal of Ambient Intelligence and Humanized Computing. pp. 1-20.

[16] Noel, M. M., Muthiah-Nakarajan, V., Amali, G. B., & Trivedi, A. S. 2021 *A new biologically inspired global optimization algorithm based on firebug reproductive swarming behaviour*. Expert Systems with Applications, pp. 115408.

[17] N. Thangadurai and Dr.R.Dhanasekaran, 2013 *A Review of Clustering based Energy Efficient Genetic Algorithms for Wireless Sensor Networks*, European Journal of Scientific Research, Vol. 101-3-4, pp. 360-371, 2013.

[18] N. Thangadurai and Dr.R.Dhanasekaran, 2013 *Energy Efficient Cluster based Routing Protocol for Wireless Sensor Networks* (International Journal of Computer Applications) Vol. 71, No. 7, pp. 43-48, 2013.

IOP Conf. Series: Earth and Environmental Science 1057 (2022) 012007 doi:10.1088/1755-1315/1057/1/012007

Preparation of soap using a steam extraction process from leaves of *Azadirachta indica*, *Ocimum basilicum*, *Hibiscus-rosa-sinensis* Flowers, *Acalypha indica*, and *Aloe barbadensis* Leaflets

C Indirani[1], K Meenambika[2], D Indhumathy[3], V S Kavinkumar[4]

[1] Department of Mathematics, Bannari Amman Institute of Technology, Sathyamangalam, Erode – 638 401, Tamil Nadu, India

[2] Department of Mathematics, Sri Shanmuga College of Engineering &Technology, Sankari, Salem, India

[3] Department of Mathematics, Sri Ramakrishna Engineering College Coimbatore, India Coimbatore, India

[4] Department of Biomedical Engineering, Bannari Amman Institute of Technology, Erode – 638 401, Tamil Nadu, India

Abstract: Soaps are used widely by humans in many aspects. Exclusively, the emergence of the SARS-Cov-2 virus made people wash their hands frequently to disinfect the virus to prevent virus infection. Soaps made of herbals possess constituents that goodness the skin as well to rejuvenate the mind and body factors. Dried leaves of Basil, Neem and Acalypha Indica, Aloe vera, and Hibiscus flower are used here to prepare homemade soap. The steam distillation process is an imperative process by which the oils are used from raw materials like dried leaves for the preparation of soaps. Extracts obtained from the herbal plants are used as additives in the preparation of soap that would be added besides the lye and other constituents. Lye preferred here is Sodium Hydroxide (NaOH) and the cast shapes utilized here in determining the soap shape are Elliptic and Rectangle. pH value, the total fatty matter is determined using respective methodologies, and the materials used in the preparation of soap are estimated using the accessible software called SoapCalc Recipe Calculator. Soaps made using the mentioned ingredients are safe and robust for cleaning the skin and hands.

Keywords: SARS-Cov-2 Virus, Herbal plants, Steam distillation. SoapCalc.

1. Introduction:

Combatting the SARS-CoV-2 virus became inevitable for all humankind over the past 2 years to survive against the fatal causing virus. Cleaning hands using soaps pave one of the ways to restrict the contagious spread of the virus. Henceforth, soaps play an indispensable role in the killing germ in which the fat globules trap the infection-causing germs and will conduit while washing with water.

Soap is a lubricating product and cleansing agent which is usually manufactured using fatty acids, containing one of the primary constituents as sodium or potassium salts. Generally, soaps are manufactured in distinct shapes and textures corresponding to their harness and their application agents. Lye is typically a concentrated alkaline solution principally consisting of sodium hydroxide and potassium hydroxide which is used in the manufacturing of soaps. Soaps are manufactured from herbal plants such as dried leaves of Neem, Hibiscus flower, Acalypha indica, Basil, and Aloe vera. By steam extraction process, instead of adding water in the manufacturing of soap, herbal constituents such as dried leaves of Neem, Hibiscus flower, Acalypha indica, Basil, and Aloe vera are preferred to make soaps.

2. Materials required and Methodologies:

Materials used in the manufacturing of bathing soap are enlisted into Lye, water, and fatty acids.

2.1. Lye: Lye used here in the production of bathing soap is Sodium Hydroxide (NaOH). Lye is crucial for the Saponification process which is the conversion of fatty acids into soaps.

2.2. Mould/Casts: Mould or cast is a type of container in which liquid or semi-solid is poured to attain the desired shapes which ultimately turned into solid soaps. Here, rectangle and oval or elliptical shaped molds are used for the preparation of soap in a home which is depicted in Figures 1 and 2.

2.3. Azadirachta indica (Neem) Leaves: A ubiquitous tree in which dry leaves are used in the treatment of skin disorders that have antioxidant agents naturally. The extract from its dried leaves is used here as one of the constituents in the preparation of soap bars [8].

2.4. Hibiscus Flower: Hibiscus which is scientifically referred to as Hibiscus rosa Sinensis is used as one of the additives for homemade soaps using its extract acquired from dried hibiscus flowers [3].

2.5. Ocimum Basilicum (Basil) Leaves: Basil leaves in general are used as one of the constituents for making soap bars. Dried leaf extract of basil leaves consists of anti-acne properties, besides, providing nourishment and moisture to the skin [9].

2.6. Acalypha Indica Leaves: Dried leaves extract of Acalypha Indica, commonly is used as an effective remedy for skin wounds and eczema sores.

2.7. Aloe barbadensis (Aloe vera) Leaves: Aloe vera gel extracted from the leaves is best known for its anti-inflammatory properties which mitigate the soreness of pain or injury caused in the skin [10].

2.8. Coconut Oil: Coconut oil is added to the preparation of soap that has cleansing properties. The type of coconut oil most commonly used by soap makers is refined, either at 76 or 92 degrees. Check with your manufacturer to see if they list refined coconut oil with words like 100% pure or with no added chemicals.

2.9. Caustic Soda: For the Saponification process, caustic soda plays a crucial factor in soap making. When flakes or beads of sodium hydroxide get delivered to a liquid, a lye solution is formed. This solution, while blended with oils or fats, will result in the chemical response referred to as saponification [7].

2.10 Water: Distilled water is preferably used in the preparation of soaps. As well, tap water also certainly utilizes tap water to make your soap bars, and many soap manufacturers do, but it should be noted that there is a difference between soap made with hard water and that used in hard water versus soap that is made with distilled water and used in distilled water [11].

The proportion of Caustic Soda: Water: Coconut Oil is 1:2:4 approximately for manufacturing soap bars.

Figure 1. Mold in the shape of Oval/Elliptical

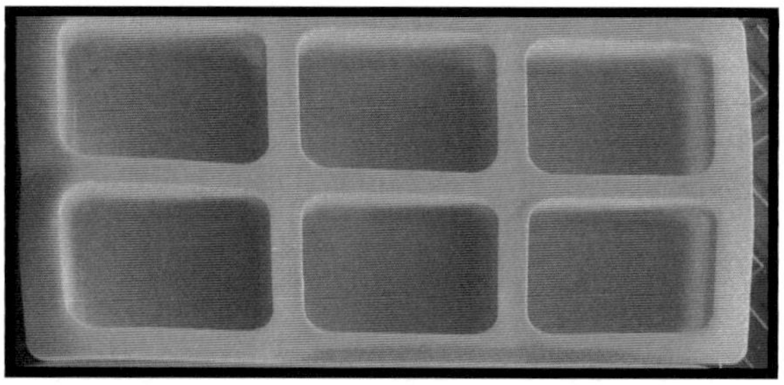

Figure 2. Mold in the shape of a Rectangle

3. Determination of pH value:

pH value refers to the degree of acidity or basicity of any solution. Typically, the pH value of the washing soap bar ranges between 9 and 10. The powder of economic soap bars (10 g) was weighed and dissolved in water. the answer was made up to 100 mL to organize 10% soap solution. The pH meter was used for calibrating a solution of pH between 8 and 10 before measuring the pH of the soap solutions [4]. Two grams of prepared soap each were dissolved in saturated mL of water and retained till the sample dissolved, and their pH values were determined with an equivalent instrument [1].

4. Steam Extraction process:

Steam distillation is one of the extensively used methods to extract the oils for manufacturing soap bars. During the process of steam distillation, the herbal plant is poured into still and pressurized steam is forced through the plant material [2]. Imagine using a steaming basket for your stove. Hot steam presses on aromatic compounds. The compound escapes from the plant material and evaporates into vapor [6]. Steam distillation is preferred for herbal extracts as the plant products can be recovered and heat-sensitive compounds can be separated whereas the fractional distillation process does not hold the mentioned properties. The herbal plant preferred here are, Basil, Neem, and Acalypha along with the Hibiscus flowers and leaflets of Aloe vera.

5. Determination of Total Fatty Matter:

Determining total fat content Total carboxylic acids (TFM) is defined because the total amount of fatty acids that are predominantly fatty acids and may be separated from the sample after cleavage with a mineral acid, predominantly HCl [1]. TFM was one of the hallmarks of the standard and texture of the soap. For this reason, TFM is usually listed on commercial soaps. during this study, TFM decided to use the tactic described in AOCS (ISO, 1975) with minor modifications. The test was performed by reacting a soap sample with acid within the presence of hot ethanol and measuring the resulting carboxylic acid. A 10-gram soap sample was mixed with 150 ml of warm neutralized ethanol and heated [5]. Soapstock has been disbanded. The dissolved solution was then filtered and therefore the residue was oven-dried at 110 ° C for 1 hour and reweighed. This can be described using the following equation 1.

$$100 - \frac{(MC+MIA)}{1.085} \tag{1}$$

where MC stands for Moisture Content and MIA stands for Matter Insoluble in Alcohol.

6. Estimation of Materials used:

Quantization and standardization of materials used for the production of soap such as lye can be estimated using the SoapCalc calculator (http://www.soapcalc.net/calc/SoapCalcWP.asp), which is depicted in Figure 3 is an open-source recipe calculator applicable for homemade soaps. Additionally, the software is used in determining the aspects such as Weight of Oils used, Water, Recipe Oil List, Soap Qualities, Fatty Acids, and so on. Hereabout, Oil weight is determined in the terms of Grams. Water is estimated as "Water as % of Oils".

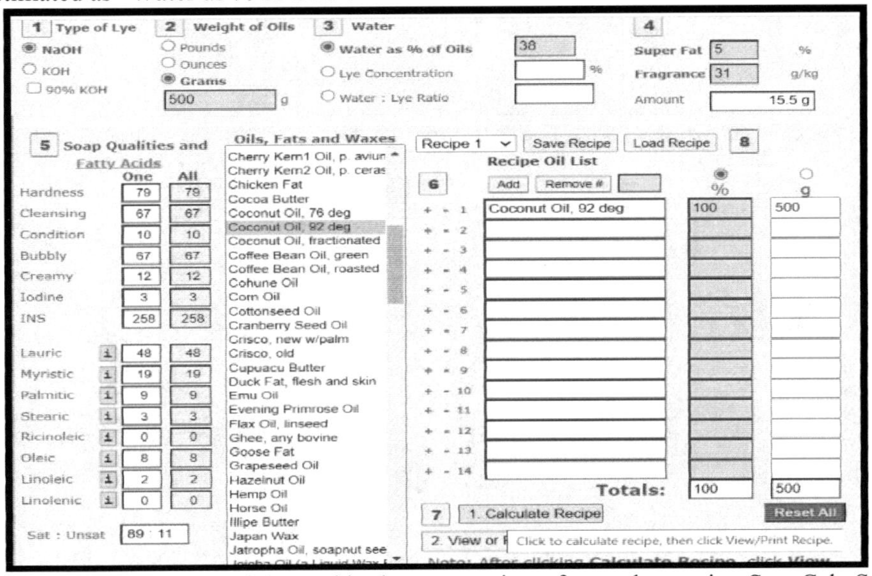

Figure 3. Estimation of Materials used in the preparation of soap done using SoapCalc Software

7. Results and Discussion:

For the preparation of soap, several aspects need to be concerned such as the type of lye used, the ratio of Water: Lye, Oil Weight, Saturation: Unsaturation Ratio, Iodine Proportion, and Super Fats. Such inclusions are enlisted as a tabulation below to discuss the constituents used for soap preparation in the home. Total weight of oil, Water as a percent of oil weight, Super fat or Discount, Concentration of Lye, and Ratio of water to lye are tabulated below in Tabulation 1. Values of Saturation: Unsaturated ratio, Iodine, and INS in Tabulation 2. Parameters like the quantity of Water, Lye (NaOH), Oils, and Soap weight before CP cure or HP cook as Pounds, Ounces, and Grams are tabulated in Tabulation 3.

Table 1. Representation of table for the values of Total oil and Water percent oil weight, Super fat, Lye Ratio, and Concentration with water

Total oil weight	500 g
Water as a percent of oil weight	38.00 vol %
Super Fat/Discount	0 g%
Lye Concentration	32.534 g%
Water: Lye Ratio	2.0737:1

Table 2: Representing a table for the values of Saturation, Iodine, INS, and Ratio Levels

Saturation: Unsaturation Ratio	89:11
Iodine	3 g
INS	258 g

Table 3. Representing the table of different parameters in Pounds, Ounces, Grams

	Pounds	**Ounces**	**Grams**
Water	0.419	6.70	190.00
Lye – NaOH	0.202	3.23	91.62
Oils	1.102	17.64	5000.00
Soap Weight before CP cure or HP cook	1.723	27.57	781.62

Figure 4

Figure 5

Figures 4, 5. Soap bars were developed using the extracts of the Neem, Basil, Hibiscus flowers, Acalypha indica, and Aloe Vera gel in different colors, textures, and shapes

8. Conclusions:

1. The motive of manufacturing this soap is to not harm the human body's epidermal tissues and to treat skin-related disorders.
2. Herbal plants used in the preparation of soap bars such as Neem, Basil, Hibiscus, Acalypha indica, and Aloe vera are easily attainable and affordable by soap manufacturers.
3. As well as to use as a solution for circumscribing the spread of the virus which may get transmitted through contact with hands.
4. Almost, soap constituents include ingredients possessing herbal properties which are safer for human skin.
5. The soap kind does not possess any fragrance factors.
6. Color and Fragrance depend on the constituents added during the manufacturing of soap.
7. Besides everything, this would be an investment for those who aspire to be an entrepreneur and own requisite knowledge regardless of domain.

9. References:

[1] Legesse Adane, *Preparation of Laundry Soap from Used Cooking Oils: Getting value out of waste*. Academic Journals, ISSN 1992-2248

[2] Md. Shafiq Alam (LM-9341) and Sandhya (LM-10052), *Extraction of Essential Oils - Need of the Hour,* Agricultural Engineering Today, **Vol. 36(1)**, 2012

[3] P. Ruban and K. Gajalakshmi, *Invitro* antibacterial activity of *Hibiscus rosa-Sinensis* flower extract against human pathogens, Asian Pac J Trop Biomed. 2012 May; **2(5)**: 399–403. DOI: 10.1016/S2221-1691(12)60064-1

[4] Jose Tarun, *Evaluation of pH of Bathing Soaps and Shampoos for Skin and Hair Care*, Indian Journal of Dermatology **59(5)**:442-4, DOI: 10.4103/0019-5154.139861

[5] K. J. Betsy, Mary Jilu, Reshma Fathima and Jaya T. Varkey, Department of Chemistry, St. Teresa's College, Ernakulam - 682 035, Kochi, Kerala, India, *Determination of Alkali Content & Total Fatty Matter in Cleansing Agents*, Asian Journal of Science and Applied Technology, ISSN 2249 - 0698 **Vol. 2** No. 1, 2013, pp.8-12

[6] Phineas Masango, *Cleaner Production of essential oils by steam distillation*, Journal of Cleaner Production **13(8)**:833-839, DOI: 10.1016/j.jclepro.2004.02.039

[7] S Rahayu, K A Pambudi, A Afifah *et.al*, *Environmentally safe technology with the conversion of used cooking oil into soap*, Journal of Physics: Conference Series, DOI: 10.1088/1742-6596/1869/1/012044

[8] Rudra Prasad Giri *et.al*, *Neem the Wonder Herb: A Short Review*, International Journal of Trend in Scientific Research and Development (IJTSRD) **Volume: 3**, Issue: 3, Mar-Apr 2019, e-ISSN: 2456 – 6470

[9] D. Lupton *et.al*, *Basil: A natural source of antioxidants and nutraceuticals*, Natural Products and Their Active Compounds on Disease Prevention (pp.463-471)

[10] Amar Surjushe, Resham Vasani, and D G Saple, *ALOE VERA: A SHORT REVIEW*, Indian J Dermatol. 2008; **53(4)**: 163–166, DOI: 10.4103/0019-5154.44785

[11] Farooq Atiku et.al, *Production of Soap Using Locally Available Alkaline Extract from Millet Stalk: A Study on Physical and Chemical Properties of Soap*, International Journal of Advanced Research in Chemical Science (IJARCS) **Volume 1**, Issue 7, September 2014, PP 1-7 ISSN 2349-039X

IOP Conf. Series: Earth and Environmental Science 1057 (2022) 012008 doi:10.1088/1755-1315/1057/1/012008

Survey on usage of single use plastic bags in Nepal

Ashish Khanal
Department of Sustainable Engineering, TERI School of Advanced Studies, New Delhi, India

E-mail: ashishkhanalk@yahoo.com

Abstract. The rapid urbanization and economic growth have caused increase of plastic products globally. The least developed countries (LDCs) have been facing severe problems with solid waste due to lack of people's awareness, lack of skilled manpower to handle waste and necessary equipment to manage in sustainable manner. Nepal, a LDC in South Asia, lacks source segregation leading the maximum plastic items reaching to the landfill site. It is even difficult for the planning authorities to make effective plans for sustainable management of plastic waste due to lack of waste generation data. Nepal has made many rules and directives to ban plastic bags till date but has failed in its implementation. This study is the first of its kind to capture the perception of youths towards single use bags in Nepal. This study was conducted among 569 Nepalese youths of age between 18-40 years. A closed ended self-administered questionnaire was distributed among the respondents and responses were collected for three weeks. The questionnaire was prepared using google form and distributed via Facebook. A snowball method was used where the respondents were asked to share the questionnaire in their network. It was found that majority of Nepalese youths have been practicing the source segregation at their houses. The use of single use plastic is very common in Nepal with an individual using at least 208 plastic bags in a year. The gender and source segregation of waste was statistically significant with p-value of 0.02 (<0.05). The main reason of using plastic bags was its easy availability. There is lack of alterative to the plastic bags in Nepal. However, the youths are ready to support the government plastic ban policy and have asked the alternatives to the plastic bags. The result has shown that the Nepalese people have negative perception towards single use plastics and are in search of alternatives. The government need to implement heavy penalty for producing plastic bags greater than 30 micron and deploy tax with shop keepers and customer if found using it.

1. Introduction

Plastic is used widely used in daily life. In spite of its major usage, it is equally important to ensure that the plastic don't affects the ecosystem chain, posing threat to human health and lives. Mainly, the micro plastics with synthetic fibres are very harmful to the living beings [1]. It is necessary to reduce the usage of plastic related items. The recycling of plastics includes process of sorting, cleaning, shredding, melting and remoulding of the materials. An individual has to conscious about their role for the environment protection. This study was done to estimate the number of plastic bags used by Nepalese which will even help the government to understand the willingness of people looking for the alternatives.

The word plastic was first coined by Leo Baekeland in 1907 by introducing first truly synthetic polymer called Bakelite [2]. Until 1970s, the paper bags were used for grocery purpose [3]. Currently huge amount of plastic is produced with an estimation of about 11% of plastic entering into the aquatic ecosystems [4] [5]. The amount of plastic production has increased over past 50 years with 335 million tonnes produced per annum in 2016 which is forecast to become 670 tonnes per annum by 2036 [6] [7].

The rapid urbanization and economic growth have caused increase of plastic products globally [8] [9]. According to a report published by the World Bank [10], around 242 million tonnes (Mt) of plastic waste was generated in 2016 which accounts for 12% of the total municipal solid waste. Also, around 5 trillion plastic bags are produced in a year globally [11] with annual usage of plastic bags per person in the world considered more than 300 [12].

Southern Asia was the top regional generator of plastic waste in 2015 accounting for 18.4 Mt per year [6]. Once the plastic reaches landfill site, the chemicals leach from plastics affects the surrounding habitats [13]. Even the burning of plastic bags because the decrease in ozone layer and causes greenhouse effect [14]. Thus, it is already late to take strict action against the usage of plastic bags. In Asia, Bangladesh and Bhutan have introduced policies prohibiting the usage of plastic bags [15]. Also plastic bags have been banned in several states of India [14]. A report from UN Environment and World Resource Institute (WRI) claimed that at least 127 countries in the world have implemented some policies related to the prohibition of plastic bags as of July 2018 [11]. It is also found that only 55 countries had adopted complete ban, eighty nine countries partially ban. Though various countries have taken measures to reduce plastic bags, yet several countries still lack any implementation strategies [16]. The incineration of plastic waste is adopted as common option for its management. There is need to environment friendly green technology [17].

Nepal is a country located in South Asia facing challenges with the management of solid waste [18] [19]. A solid waste management survey [20] conducted in Nepal found that the average daily waste generation of Nepal is 1435 tons per day with per capita waste generation per day as 0.17 kg. It comprises of 12% of plastic related materials with maximum reaching to the landfill site due to less recovery of the recyclables [21]. A recent study [22] found that the amount of solid waste generated per day in Nepal has increased to 1653 tons. The amount of waste generation in Nepal is expected to rise rapidly in next five year with per capita MSW generation expected to become 0.7 kg/day by 2025 [10].

There is lack of source segregation, poor waste collection service and open dumped landfill in Nepal [23]. The single use plastic bags have been used in Nepal from 1990s [24]. The plastic bags are provided free of charge by the shops in Nepal and in other parts of the world as well leading to the excessive use [25]. Nepal has made many rules and directives to ban plastic bags (table 1) till date but has failed in its implementation [26] [27]. Nepal has made several attempts to ban plastic bags starting from Hetauda in 1998. Single use plastic bags thinner than 30 microns have been banned in different cities of Nepal including Kathmandu, Hetauda, Pokhara, Illam, Palpa and even in the Mount Everest region. Nepal also issued Plastic Bags Regulation and Control Directive in 2011 which prohibited the use, import, store and sale of plastic bags less than 30 microns. Though plastic ban has been announced in several cities of Nepal, but there are no bans on plastic bags manufacturers.

The plastic is used in electronics, furniture, machineries, packaging materials [28] and widely as the household items in our daily life [29]. The most common single use plastic are used as plastic bags, plastic bottles, disposable cups and food packaging items [29]. The plastic is used commonly in form of plastic bags in household settings [29]. But, the single use plastic bags are the most harmful [30]. The unmanaged or scattered plastic bags can cause socio-ecological impacts affecting the natural beauty and causing visual pollution [31].

The recovery rate of plastic is very low leading the most of the plastic waste reaching to the ocean, disposed at landfills or burned in incinerators [8]. The usage of plastic bags have negative impact on the environment [32] [33] [8] [7] and very difficult to decompose naturally [34]. Though there are several negative impacts of plastic bags, people still prefer to use it [35] [36]. Researchers agree that the main reasons for widely usage of plastic bags are its durability, low weight, low cost and easy to carry [37] [38] [39]. It is necessary that a product to be designed in such a way that it causes zero to least environmental impacts [40]. There are several benefits of recycling which includes protecting the environment, cut down the transport and reducing the amount of waste reaching to the landfill site [14] [41]. However, only 16 countries in the world have rules regarding the use of reusable bags or alternatives to plastic bags [11]. The knowledge, attitudes, and behaviours of an individual toward plastic waste management need to change in order to reduce the plastic waste [8]. The knowledge, practices and usage of plastic bag also varies between one people to other. There is no study conducted till date focusing on the perception of Nepalese youths towards plastic bags. Thus, this study was conducted to find the usage of plastic bags among youths of Nepal and their interest for shift towards the alternatives.

Table 1. History of plastic ban in Nepal.

Year	Progress
1995	Hetauda became first city in Nepal to ban plastic bags
2002	Supreme court directed the Nepal government to ban plastic bags
2011	Nepal issued directive to ban thin plastic bags
2013	Kathmandu, the capital city declared complete ban of plastic bags
2015	Ministry of Science, Technology and Environment (MoSTE) banned plastic bags; published in official gazette about ban of plastic bags smaller than 20 inches in width, 35 inches in length and less than 40 micron
2018	Nepal decided to ban plastic bags thinner than 40 micron
2021	Nepal announced complete ban on the production, import, sale and usage of plastic bags thinner than 40 microns

2. Materials and methods

2.1. Population and sample

The study was conducted among youths of Nepal with age above 18 years and below 40 years. According to the Central Bureau of Statistics, the youth's population of Nepal is 12,210,457. The sample size was calculated using the standard formula [42] at 95% confidence level with 4.1% margin of error.

$$n = \frac{z^2 * N * p(1-p)}{Nd^2 + z^2 * p(1-p)}$$

Where:

z 95% confidence level (1.96)
N total population size
p estimated population proportion
d margin of error
n sample size

The above formula gave a sample size of 569. According to Krejcie and Morgan table [43], a sample of 384 is sufficient to conduct a study with population of 1,000,000 or above.

2.2. Data collection

A closed ended self-administered questionnaire was distributed among the respondents and responses were collected for three weeks. The questionnaire was prepared using google form and distributed via Facebook. The youths were purposively chosen for this study with the option to be filled by people of age between 18-40 years old. A snowball method was used where the respondents were asked to share the questionnaire in their network. In three weeks period, the response was received from 569 respondents. The survey questionnaire included information on socio-demographic status, usage of plastic bags and their interest shifting towards the alternative options. The obtained information was imported in Statistical Package for the Social Sciences (SPSS) software to analyse the data.

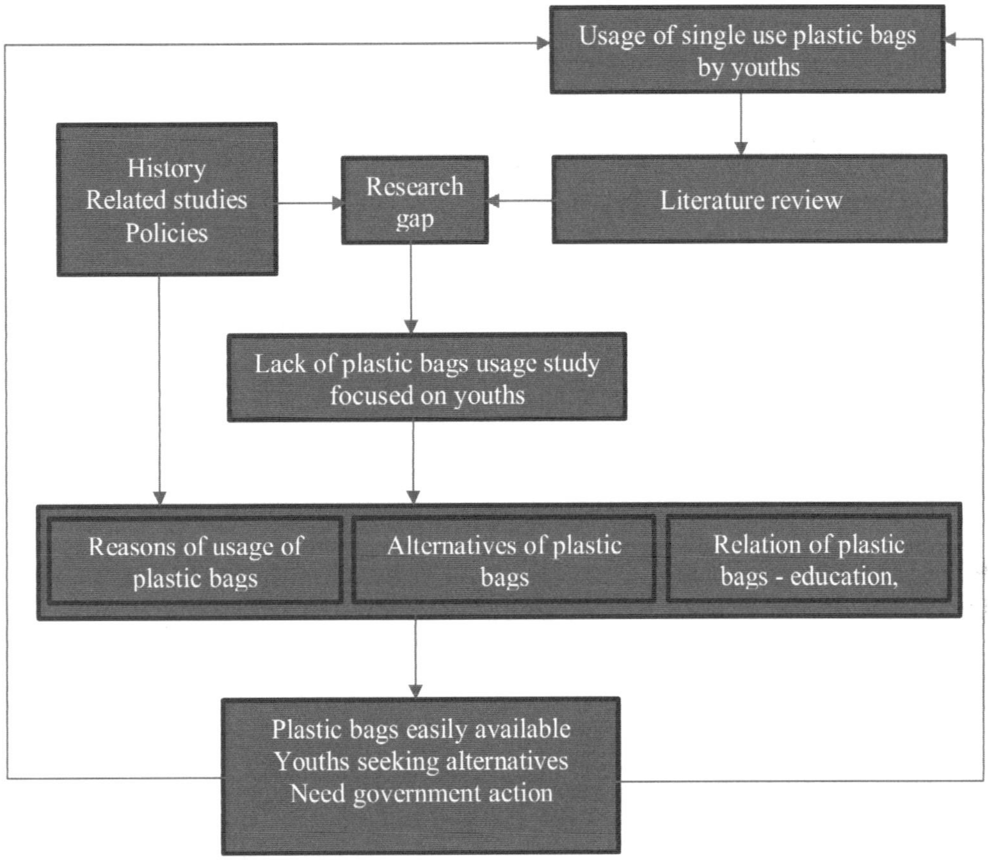

Figure 1. Conceptual framework of the study on usage of plastic bags by youths

3. Results and discussion

3.1. Demographic details

A total of 569 people participated in this study. The participation of female (56.1%) respondents was higher compared to males (43.9%). As the study was conducted using google survey, all respondents were literate with majority (69.6%) of respondents completing their bachelor degree.

Table 2. Demographic details of the respondents.

Characteristics	n=569	%
Gender		
Male	250	43.9
Female	319	56.1
Education		
Lower Secondary	2	0.4
Secondary	3	0.5
Higher Secondary	60	10.5
Bachelor	396	69.6
Graduate	92	16.2
Post graduate and above	16	2.8

3.2. Usage of plastic bags

It was found that majority (80.1%) of respondents have been practicing source segregation at their house (table 3). A question was asked as how often the respondents carry bags from house at the time of going for shopping, 35.9% claimed that they always take bags with them. Another question was asked as how many plastic reaches their home in a week (both respondents and their parents), majority (41.1%) of respondents said the number of plastic bags was between 4-6. It is similar to a study [44] conducted in Nigeria which found the respondents using atleast seven plastic bags in a week. Also a study [45] conducted in Bangladesh found that about 35% consumers use 5-10 plastic bags per week. Even in India, the average number of plastic bags used in a week is 3 while in China, an individual uses 21 plastic bags in a week [46]. Around 19.0% respondents in Turkey are taking 5-6 plastic bags in week [12]. Based on the finding, atleast 208 plastic bags are used annually by an individual in Nepal (taking 1 year equals 52 weeks and an individual using at least 4 bags in a week). But this study is limited to the people who have access to the social media. The consumption of plastic bags in rural areas of Nepal is very less compared to cities.

A question was related to the reasons of using plastic bags by respondents where multiple responses were recorded. 80.8% respondents have been using the plastic bags because it is easily available. The result is higher than the study conducted in Ethiopia claiming 69.1% [47], 40% in South Africa [48] and 33.5% in Pakistan [49]. Even studies conducted in Kenya [50], Uganda [51], Malaysia [34] and Bangladesh [45] have found easy availability, low price and lack of alternatives as the main reasons of using plastic bags.

Around 98.6% respondents of this study are ready to quit using plastic bags and start using other alternatives if provided by the government. A study [51] conducted in Uganda found that 96.4% respondents had shown interest for alternative to plastic bags. Similarly, a study conducted in Botswana found that the majority of the respondents (80.7%) are willing to accept a shift to eco-friendly alternatives for plastic bags [33]. Even a study in United States found that 91% respondents prefer to use alternatives of plastic bags [52]. It means the youth of Nepal are looking for the alternatives and really worried about the impact of using the plastic bags. A study conducted among retailers and consumers in Nepal concluded that a partial ban of single-use plastic bags isn't sufficient to change the behaviour [24]. There is need of proper monitoring mechanism so that every stakeholders engage under a system. The ban of single use plastic bags also helps to meet sustainability goals with alternatives leaving smaller footprint on the earth.

Table 3. Respondents view on the usage of plastic bags.

Characteristics	n=569	%
Waste segregation		
Yes	456	80.1
No	113	19.9
How often carry bags from house		
Always	204	35.9
Sometimes	117	20.6
Often	144	25.3
Rarely	75	13.2
Never	29	5.1
How many plastic bags		
0	6	1.1
1-3	177	31.1
4-6	234	41.1
7-9	80	14.1
10 and above	72	12.7
Reason of plastic bags used		
Cheap	301	52.9
Easily available	460	80.8

Light weight	166	29.2
No other alternative	171	30.1
Ready to quit if given alternative		
Yes	561	98.6
No	8	1.4

The gender and source segregation of waste was statistically significant with p-value of 0.02 (<0.05). The p-value (0.39) for source segregation with education was found be more than 0.05 which shows no association between the attributes (table 4). The chi-square significance (p) showed no association between gender and respondents interest to quit the usage of plastic bags. However the education with interest in quitting using plastic bag was statistically significant with p-value of 0.001 (<0.05). The education campaign can help to reduce the plastic bags at source [16]. So it is necessary that the government mobilize environmental campaigns in order to educate people about the problems associate with the improper disposal of plastic bags [33] [29].

Table 4. Factors attributed for usage of plastic bags.

Characteristics			Yes	No	p-value
Source segregation	Education	Lower Secondary	1 (0.2%)	1 (0.9%)	0.39
		Secondary	2 (0.4%)	1 (0.9%)	
		Higher Secondary	43 (9.4%)	17 (15.0%)	
		Bachelor	320 (70.2%)	76 (67.3%)	
		Graduate	76 (16.7%)	16 (14.2%)	
		Post graduate and above	14 (3.1%)	2 (1.8%)	
	Gender	Male	186 (40.8%)	64 (56.6%)	0.02
		Female	270 (59.2%)	49 (43.4%)	
Quit plastic bags	Education	Lower Secondary	1 (12.5%)	1 (0.2%)	0.001
		Secondary	0	3 (0.5%)	
		Higher Secondary	0	60 (10.7%)	
		Bachelor	6 (75.0%)	390 (69.5%)	
		Graduate	1 (12.5%)	91 (16.2%)	
		Post graduate and above	0	16 (2.9%)	
	Gender	Male	244 (43.5%)	6 (75.0%)	0.07
		Female	317 (56.5%)	2 (25.0%)	

4. Conclusion and Recommendation

It was found that majority of Nepalese youths have been practicing the source segregation at their houses. The use of single use plastic is very common in Nepal with an individual using at least 208 plastic bags in a year. The main reason of plastic bags being widely used in Nepal was its easy availability. The gender and source segregation of waste was statistically significant but no association was found between the two attributes. Youths are ready to support the government plastic ban policy and have asked the alternatives to the plastic bags. Though the plastic bags are convenient for consumer, it causes extreme harmful to the environment. The alternatives to plastic bags could be cotton bags, paper bags, compostable bags or jute bags. Nepal Government need to implement heavy penalty for producing plastic bags greater than 30 micron and deploy tax with shop keepers and customer for using it. In fact, if the factories are banned from producing single use plastic bags greater than 30 micron, it will barely reach in the hand of people. However, there is chance of its usage mainly in the region bordered with India (as Nepal and India shares open border), so a strict monitoring mechanism is required.

Acknowledgments
The author is very grateful to Mr. Prasuj Mainlai and Ms. Suja Giri for sharing the google form within their circle. The author is equally thankful for all the respondents who participated in this study.

References

[1] Shanmugam K 2022 Spray Coated Cellulose Nanofiber Laminates on the Paper to Enhance its Barrier and Mechanical Properties *J. Sustain. Environ. Manag.* **1** 10–7

[2] Brydson J A 1999 *Plastics Materials 7th Edition* vol 7

[3] Jalil M A, Mian M N and Rahman M K 2013 Using Plastic Bags and Its Damaging Impact on Environment and Agriculture: An Alternative Proposal *Int. J. Learn. Dev.* **3** 1

[4] Borrelle S B, Ringma J, Law K L, Monnahan C C, Lebreton L, McGivern A, Murphy E, Jambeck J, Leonard G H, Hilleary M A, Eriksen M, Possingham H P, De Frond H, Gerber L R, Polidoro B, Tahir A, Bernard M, Mallos N, Barnes M and Rochman C M 2020 Predicted growth in plastic waste exceeds efforts to mitigate plastic pollution *Science (80-.).* **369** 1515–8

[5] Modak A and Basu S 2022 Microplastic- An imposing commination to the aquatic ecosystem and its removal strategies in Wastewater Treatment Plants: A Systematic Review *J. Sustain. Environ. Manag.* **1**

[6] Lebreton L and Andrady A 2019 Future scenarios of global plastic waste generation and disposal *Palgrave Commun.* **5** 6

[7] Drzyzga O and Prieto A 2019 Plastic waste management, a matter for the 'community' *Microb. Biotechnol.* **12** 66–8

[8] Chow C-F, So W-M W, Cheung T-Y and Yeung S-K D 2017 Plastic Waste Problem and Education for Plastic Waste Management *Emerging Practices in Scholarship of Learning and Teaching in a Digital Era* (Singapore: Springer Singapore) pp 125–40

[9] Khanal A, Giri S and Mainali P 2022 Practices of Source Segregation of Household Solid Waste by Youths of Nepal *SSRN Electron. J.*

[10] Kaza S, Yao L C, Bhada-Tata P and Van Woerden F 2018 *What a Waste 2.0: A Global Snapshot of Solid Waste Management to 2050*

[11] Excell C 2019 127 Countries Now Regulate Plastic Bags. Why Aren't We Seeing Less Pollution?

[12] Arı E and Yılmaz V 2017 Consumer attitudes on the use of plastic and cloth bags *Environ. Dev. Sustain.* **19** 1219–34

[13] Rochman C M, Browne M A, Halpern B S, Hentschel B T, Hoh E, Karapanagioti H K, Rios-Mendoza L M, Takada H, Teh S and Thompson R C 2013 Classify plastic waste as hazardous *Nature* **494** 169–71

[14] Moh Y C and Abd Manaf L 2014 Overview of household solid waste recycling policy status and challenges in Malaysia *Resour. Conserv. Recycl.* **82** 50–61

[15] Gupta K 2011 *Consumer Responses to Incentive to Reduce Plastic Bag Use: Evidence from a Field Experiment in Urban India* (The South Asian Network for Development and Environmental Economics)

[16] Xanthos D and Walker T R 2017 International policies to reduce plastic marine pollution from single-use plastics (plastic bags and microbeads): A review *Mar. Pollut. Bull.* **118** 17–26

[17] Saxena A and Srivastava A 2022 Industry Application of Green Manufacturing: A Critical Review *J. Sustain. Environ. Manag.* **1** 32–45

[18] Khanal A, Sondhi A and Giri S 2021 Use of personal protective equipment among waste workers of Sisdol landfill site of Nepal *Int. J. Occup. Saf. Heal.* **11** 158–64

[19] Giri S 2021 Integrated solid waste management: A case study of a hotel in Kathmandu, Nepal *Int. J. Multidiscip. Res.* **7** 264–8

[20] ADB 2013 *Solid Waste Management in Nepal: Current Status and Policy Recommendations*

[21] Khanal A and Giri S 2016 Possibility of Community Based Solid Waste Management Project in Chuchepati, Kathmandu *J. Basic Appl. Eng. Res.* **3** 6–8

[22] CBS 2021 *Waste Management Baseline Survey of Nepal 2020* (Kathmandu)

[23] Khanal A 2021 Livelihood Status of Itinerant Waste Buyers in Kathmandu *Appl. Ecol. Environ. Sci.* **9** 537–40

[24] Bharadwaj B, Baland J M and Nepal M 2020 What makes a ban on plastic bags effective? The case of Nepal *Environ. Dev. Econ.* **25** 95–114

[25] Ayalon O, Goldrath T, Rosenthal G and Grossman M 2009 Reduction of plastic carrier bag use: An analysis of alternatives in Israel *Waste Manag.* **29** 2025–32

[26] Bharadwaj B 2016 *Plastic Bag Ban in Nepal: Enforcement and Effectiveness* (Kathmandu)

[27] Shrestha H S, Poudel R and Bhandari D P 2016 Plastic Bags and Control Initiatives in Nepal *J. Environ. Sci.* **2**

[28] Zheng Y, Yanful E K and Bassi A S 2005 A Review of Plastic Waste Biodegradation *Crit. Rev. Biotechnol.* **25** 243–50

[29] Khan F, Ahmed W and Najmi A 2019 Understanding consumers' behavior intentions towards dealing with the plastic waste: Perspective of a developing country *Resour. Conserv. Recycl.* **142** 49–58

[30] Knoblauch D, Mederake L and Stein U 2018 Developing Countries in the Lead—What Drives the Diffusion of Plastic Bag Policies? *Sustainability* **10** 1994

[31] Zambrano-Monserrate M A and Alejandra Ruano M 2020 Do you need a bag? Analyzing the consumption behavior of plastic bags of households in Ecuador *Resour. Conserv. Recycl.* **152**

104489

[32] Asmuni S, Hussin N B, Khalili J M and Zain Z M 2015 Public Participation and Effectiveness of the no Plastic Bag Day Program in Malaysia *Procedia - Soc. Behav. Sci.* **168** 328–40

[33] Madigele P K, Mogomotsi G E J and Kolobe M 2017 Consumer willingness to pay for plastic bags levy and willingness to accept eco-friendly alternatives in Botswana *Chinese J. Popul. Resour. Environ.* **15** 255–61

[34] Jayaraman K, Haron H, Sung G B and Lin S K 2011 Consumer reflections on the usage of plastic bags to parcel hot edible items: an empirical study in Malaysia *J. Clean. Prod.* **19** 1527–35

[35] Sun Y, Wang S, Li J, Zhao D and Fan J 2017 Understanding consumers' intention to use plastic bags: using an extended theory of planned behaviour model *Nat. Hazards* **89** 1327–42

[36] Chikaire J U, Ajaero J O and Atoma C N 2022 Socio-economic Effects of Covid-19 Pandemic on Rural Farm Families' Well-Being and Food Systems in Imo State, Nigeria *J. Sustain. Environ. Manag.* **1** 18–21

[37] Hopewell J, Dvorak R and Kosior E 2009 Plastics recycling: challenges and opportunities *Philos. Trans. R. Soc. B Biol. Sci.* **364** 2115–26

[38] Sharma B K, Moser B R, Vermillion K E, Doll K M and Rajagopalan N 2014 Production, characterization and fuel properties of alternative diesel fuel from pyrolysis of waste plastic grocery bags *Fuel Process. Technol.* **122** 79–90

[39] Wong S L, Ngadi N, Abdullah T A T and Inuwa I M 2015 Current state and future prospects of plastic waste as source of fuel: A review *Renew. Sustain. Energy Rev.* **50** 1167–80

[40] Khanal A 2021 Sustainable Living in Schools: A Study of Vajra Academy, Lalitpur *Curr. World Environ.* **16** 472–6

[41] Gumel D Y 2022 Assessing Climate Change Vulnerability: A Conceptual and Theoretical Review *J. Sustain. Environ. Manag.* **1** 22–31

[42] Arkin H and Colton R R 2007 Tables for statisticians. *Tables For Statisticians.*

[43] Krejcie R V. and Morgan D W 2016 Determining Sample Size for Research Activities: *http://dx.doi.org/10.1177/001316447003000308* **30** 607–10

[44] Iheukwumere S O, Nkwocha K F and Tonnie-Okoye N 2020 Stemming Plastic Bag Pollution In Anambra State: Willingness Of The Public To Accept Alternative Bags *coou African J. Environ. Res.* **2** 17–32

[45] Uddin M, Hasan M, Sahadat H and Khan B 2018 People perceptions about using polythene bag and its impact on environment at Mymensingh in Bangladesh *Int. J. Nat. Soc. Sci.* **5** 37–43

[46] Muthu S S, Li Y, Hu J Y and Mok P Y 2011 Carbon footprint of shopping (grocery) bags in China, Hong Kong and India *Atmos. Environ.* **45** 469–75

[47] Adane L and Muleta D 2011 Survey on the usage of plastic bags, their disposal and adverse impacts on environment: A case study in Jimma City, Southwestern Ethiopia *J. Toxicol. Environ. Heal. Sci.* **3** 234–48

[48] O'Brien J and Thondhlana G 2019 Plastic bag use in South Africa: Perceptions, practices and potential intervention strategies *Waste Manag.* **84** 320–8

[49] Ahsan M U, Nasir M and Abbas J 2020 Examining the Causes of Plastic Bags Usages and Public Perception about its Effects on the Natural Environment *Int. J. Acad. Res. Bus. Soc. Sci.* **10** 80–96

[50] Njeru J 2006 The urban political ecology of plastic bag waste problem in Nairobi, Kenya *Geoforum* **37** 1046–58

[51] Mugisha J and Diiro G 2015 Households' Responsiveness to Government Ban on Polythene Carrier Bags in Uganda *J. Agric. Environ. Sci.* **4** 2334–412

[52] Camann A, Dragsbaek K, Krol S, Sandgren J and Song D 2010 *Properties, Recycling and Alternatives to PE Bags* (Worcester Polytechnic Institute)

Applications for Green Technology Practices in Analyzing the Edge Detection Techniques on Digital Image Processing by using MATLAB

R. Ravikumar[1], V. Arulmozhi[2]
[1]Research Scholar, Department of Research Computer Science, Tirupur Kumaran College for Women, Tirupur, Tamilnadu, India
[2]Associate Professor, Department of Research Computer Science, Tirupur Kumaran College for Women, Tirupur,
Tamilnadu, India

E-mail: rravigrat@gmail.com, drarultkc@gmail.com

Abstract. Increasing awareness and growing focus on environmental issues such as climate change, energy use, and the loss of non-renewable resources have created a great need for research that provides possible solutions to these problems. Environmental science and engineering research continue to play an important role in revealing new "green" future opportunities. Digital Image Processing plays a major role in contributing to its relevant research works under these Environmental and Green Technologies. By applying tools related to Environment technology this Digital Image Processing produces numerous applications. In such cases, Edges provide an important dynamic change in digital image processing. It is a process of finding and finding edges with sharp changes made by the power of its images. This is the best way to get an accurate edge in good shape and also different edge finders will work in better different conditions. Different methods of edge detection were compared comparatively to determine which method was most suitable for obtaining edges without any noise conditions. Here, the three types of edges have horizontal, vertical and diagonal edges. In this paper, we review the different editing methods and compare different identification methods, advantages, and disadvantages. Implementation is done using MATLAB. A sudden change of discontinuity is evident in the image.

Keywords: Edge Detection, Sobel Operator, Prewitt's Operator, Robert's Operator, Log Operator, Canny Operator, Zero-Crossing Operator.

1. Introduction

Environmental engineering is a field of specialized engineering that covers a wide range of scientific topics such as Chemistry, Biology, Ecology, Geology, Hydraulics, Hydrology, Microbiology, Computer science, and Mathematics to create solutions that will protect and enhance biodiversity and improve biological quality. nature. Environmental engineering is the application of scientific and engineering principles to the development and preservation of the environment to protect human health, to protect beneficial ecosystems, and improving the ecological development of human life. Environmental engineers are studying the effects of technological advances on the environment, addressing local and global environmental issues such as acid rain, global warming, ozone depletion, water pollution, and air pollution from automotive machinery and industry resources. Environmental engineering is the name of the work done since the beginning of civilization, as people learn to adapt and control environmental conditions to meet needs. As people realize that their health is related to the quality of the environment in which they live, they formulate plans to improve it. Many universities offer environmental engineering programs through the department of computer engineering or synthetic engineering and include electrical projects to improve and standardize environmental conditions. Environmental engineers in the field of computer engineering often focus on Machine Learning, AI and Expert Systems, and the Design of Digital Image Processing. Local engineers in the Digital Graphics Program often focus on environmental innovation, advanced tools and application technologies, and their appropriate processes.

Edge detection is the name of a set of mathematical methods that aim to identify points in a digital image where the brightness of the image can change sharply or more officially, rather than continuously.

Any points where the light of the image changes sharply are usually arranged into a set of curved line segments into so-called edges. The same problem of getting pauses in the 1D signal is known as step detection and the problem of detecting signal failure over time is known as detection of changes in image processing. Edge detection is an important basic tool in image processing when machine vision and computer vision, especially in the areas of feature detection and feature removal.

Digital Image is made up of a limited number of components, each with a specific location or location and value. These components are specified as image elements, image elements, and pixels. Image processing is any type of signal processing in which an image is inserted, such as an image, and the output of the image processing may be that the image or, set of symbols or parameters associated with the Edge of the image can also be defined as binary. images like black pixels with one white neighbour nearby. The edges include a large amount of important information about the image.

Changes in pixel density define the parameters of the objects in the image. Feature detection and feature extraction are key areas for image processing, where Edge detection is an important tool. The image edge acquisition adverts from the edges of the image by detecting the maximum intensity difference in pixels. This action finds the outline of the object and the boundaries between the objects and the background of the image.

Detection of image edges can help to split images, compress data, image reconstruction, and more. Variables involved in the selection of the edge detector include the edge shape, sound area, and edge structure. Edge detection is difficult for sound images as both sound and edges incorporate the essence of high frequency. Attempts to reduce the effect of noise are blurred and distorted.

Edge detection is mainly used to extract information about an image e.g. sharpen and enhance the image, the location of the object in the image, its shape, size.

This paper organizes as follows: Section II concentrates on Literature Review, Section III clarifies the problem statements in various edge detection techniques, Section IV categorizes the different methods in edge detection techniques and Section V shows the results and discussions, Section VI finally, conclude this paper.

2. Literature Review

Mohd. Aquib Ansari et al, suggested that the discovery of the Edge could also be a way of finding the occurrence of edges and their location formed by the sharp and rapid variation of image stabilization. The purpose of the edge acquisition is to obtain the information of the object form and therefore the reflection within the image. Here different edge detection techniques are used to identify the sides during the image and thus comparative analysis is also performed between these techniques. [1].

Saket Bhardwaj et al, have suggested that in many computer-assisted vision applications this discovery plays a major role. It greatly reduces data to unwanted information and provides useful information to the image. Here are some of the problems such as false positive detection that has occurred due to noise and the lack of low boundary variability. And this paper compared the various machines on the edge to identify which edge detector provides performance with the best results [2].

K. Bala Krishnan et al, point out that in many programs these digital images play important roles such as image retrieval, transmission, robot vision, remote sensors, and pattern detection. And in these digital image processing applications, image classification is an important part because immature images are captured with a digital camera with non-essential background scenes or sound. Here sound removal is done using sound extraction techniques such as filtering and enhancing images. The main reason for these methods is to limit false detection at the edges, local edge-making, and calculation time [3].

Manzoor Ahmed Hashmi et al, have suggested that in the work of finding a horoscope for marine photography, it is very important. To reduce the cost of calculation, double image separation is divided into two regions for obtaining objects and identifying and tracking objects. Here, the survey focuses on an image-based horizontal finding method used to determine the location of a ship and its sets of test data and final results [4].

Ramnarayan et al, suggested that in processing the gray image the method of obtaining a canny edge was widely used. Using a central filter of algorithms with a quaternion weight, the vector calculation Sobel gradient provides test results in performance beyond other methods to obtain image edge. The method presented here provides excellent accuracy using a two-dimensional edge detector with low calculation costs [5].

Diplaxmi R. Waghule et al. Comparative evaluation of different edge acquisition methods make it easier to determine which edge acquisition method is appropriate for image classification [6].

Ghassan Mahmoud Hussien Amer et al argue that the acquisition of the Edge is a form of the image separation process that determines the presence of a foot or a line in an image and defines it appropriately. The main purpose of edge acquisition is to simplify image data to reduce the amount of information to be processed. A challenge is being made to review the introduction of the most commonly used cutting-edge image separation methods and the differences in these processes are being explored using MATLAB software to use edges to determine compatibility [7].

Shubhashree Savant et al, have suggested that the edges of the image reflect the boundaries and regions of the image. The edges of the images are areas with a strong brightness - jumping sharply from one pixel to the next. Edge detection is one of the most important functions in image processing and image analysis. The edge acquisition process significantly reduces the amount of data and filters useless data while maintaining important structures during image processing [8].

Muthukrishnan et al, also point out that image classification is an important step in image analysis. Separation separates an image into parts or objects. The extent to which the division is made depends on the issue being resolved. Here, an attempt is made to review a variety of commonly used edge-finding methods for image classification, and the effectiveness of such techniques is done using imagery using the MATLAB software [9].

Khalid Alshalfan et al, suggested that Edge discovery is one of the most important digital image processing methods and algorithms responsible for discovering the most important features and structures of digital imagery. A key feature of this paper is the launch and calculation of image output. The main objective of this paper is to detect the sharpening of the edge by extracting the most impressive pixels of objects in the image that may be the boundaries of an object, which may lead to a sudden change in the appearance of the image [10].

The discovery of Tasnuva Tasneem et al exposed sting may also be an important area of Image Processing that can be used for a variety of imagery including medical photography. Here, a detailed analysis based on Edge Discovery of images for various comparisons was performed. Results will be obtained by extending the histogram [11].

Satbir Kaur et al, have suggested that the Edge is the best image. The edges are often defined as the boundary between the regions in the middle of the image. The main objective is to review the Bite Acquisition process supported by various Strategies such as Prewitt, Sobel, Canny, Roberts, and LoG [12].

Tamanna Sahoo et al, have revealed that the Edge Detection Technique could be a Step into Step Process for a few Image Processing Algorithms such as image enhancement, image classification, tracking, and image or video encoding. The models used here are MATLAB and SIMULINK for Sobel, Robert, Prewitt, LoG, and their publications [13].

Manjula KA et al, suggested that Digital Image Processing could also be a way to use computer Algorithms to perform specific tasks on an image. The most common methods found here are image classification and Edge-based classification. Image Separation should separate objects and thus Acquisition of biting is considered an important tool for Image Separation [14].

J. S. Owotogbe et al, suggested that the edges be known as an important change in the firmness of Digital Images. Here, the various Edge Discovery Methods with their advantages and disadvantages were represented by the MATLAB functions [15].

Divyang Goswami et al, have revealed that the edge can also be described as the gaggle of connected pixels forming a boundary between two random regions. Here, many Edge Detection Techniques perform various features in the lower chances of finding false edges [16].

Farah Saad Al-Mukhtar et al, suggested that the acquisition of the Edge is one of the methods of image enhancement used to extract important features from the edges of the image (e.g., corners, lines, curves). The algorithm is used using the MATLAB 2010 language code and because the VHDL language will affect the use of the FPGA device, which was genre (Xilinx XC3S500E Spartan3E), and was used in 8-bit gray image data, the algorithm detection is generated Robinson edge. using pixel windows (3×3 windows) to calculate output, make comparisons between the resulting image in MATLAB and VHDL by calculating peak signal-to-noise (PNSR), Root Mean Square Error (RMSE) as well as linking between images resulting from MATLAB and VHDL [17].

Xin Baiet al proposed that bridges be exposed to many external loads, such as from tra_c, winds, floods, and earthquakes. Monitoring bridges are becoming increasingly important due to safety concerns. There are some specific measurement methods such as the use of linear variable di_erential transformer (LVDT) and laser-based migration sensors. The visual detection system for monitoring the removal of the structure is based on enhanced edge detection and subpixel modes are suggested. The performance of the proposed system is first verified by MTS testing. The field test is then applied to the sign, and thus the proposed view system is used to provide the time history and natural frequencies of the road sign immediately. Finally, the use of such a system for the detection of damage during the iron ore is explored [18].

Xiaofeng Li et al suggested that the cancer screening algorithm support in-depth study. therefore, in providing a basis for analyzing additional cancer research data, pain diagnosis is eliminated with the reconstruction of the cancer image. The algorithm first performed the reconstruction of three-dimensional cancer images and extracted fine features of cancer cells. On this basis, it deciphered elements of the color structure of the cancer images and used in-depth reading technology to discover and complete the critical information of the cancer images. The test results show that the performance of the proposed algorithm is high, which provides some reference to the long-term development of the medical field [19].

Miroslav Hagara et al propose tons of articles on the implementation of the FPGA for the acquisition of the edge in gray images and one on the implementation of the FPGA for the acquisition of the color edge. Color detection is rare because your calculation requirements are 3 times greater than the gray image and 90 percent of the edges are almost identical in color and color images so obviously, you do not need to process a color image., only in some special cases [20].

3. Problem Statement

The edge-finding method has received a lot of attention in recent years due to its widespread use in several fields. Edge Discovery is one of those challenging issues and to date, no strategy offers a solid solution to all or any situation or any of the various situations and applications that may encounter a way to find the edge.

Border detection is assessed by local domains and frequency. The performance performed by the local background is in direct pixels of the image and is understood as an operator-based approach. In addition, this method is tested according to the first and second methods. two orders have different operating modes, the first system has a positive output of the image, while the second has a second output of the image gradients, and together they operate on the image gradient, but with different commands.

4. Methods for Edge Detection Techniques

Edge Detection represents the most important part of Digital images. The display of the edges of the images has reduced the amount of information to be processed. To get the best shape for the edges these edge trimming techniques make an important tool for image classification.

Edge acquisition strategies are divided into two main categories:

- Grading-based edge acquisition that supports first-order acquisition.
- Gaussian-based edge acquisition supporting the second-order acquisition.

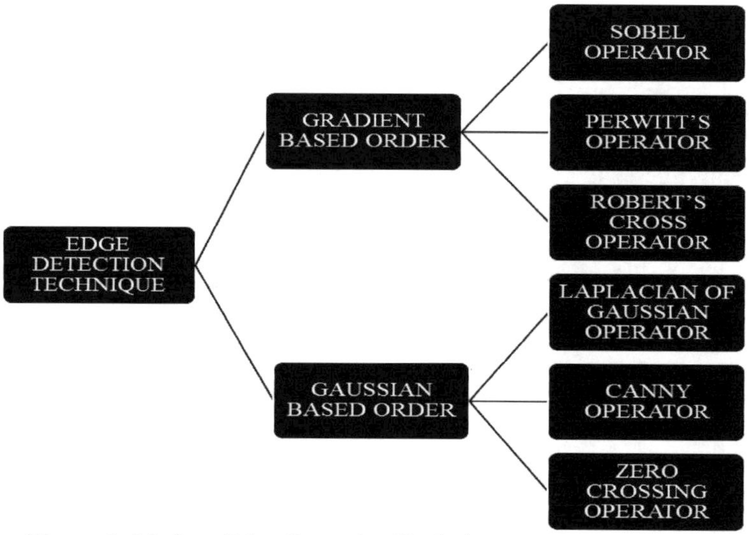

Figure 1. Various Edge Detection Techniques

Algorithm:

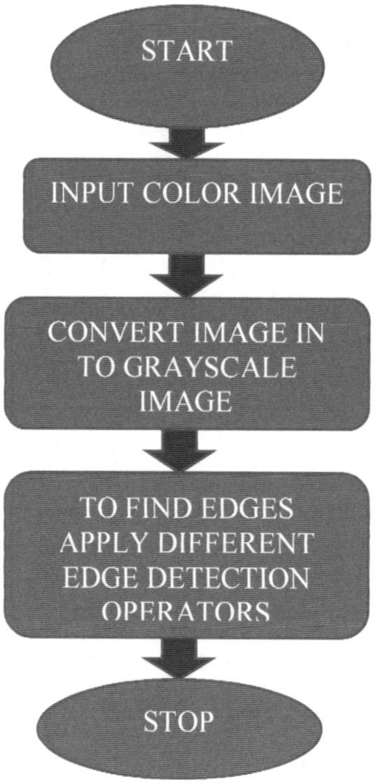

Figure 2. Flowchart of Edge Detection Techniques

Input: Sample Image Folder.
Output: Predicted Edges.
Step 1: Read the installation image folder.
Step 2: Convert RGB to gray images.

Step 3: Use the Sobel method.
Step 4: Apply the Prewitt method.
Step 5: Apply Robert's Cross Method.
Step6: Install the Gaussian Laplacian (LoG) method.
Step 7: Use the Canny method.
Step 8: Use the Zero Crossing method.
Step 9: Show the result of the Sobel method.
Step 10: Show the effect of Prewitt's method.
Step 11: Show the effect of Robert's Cross method.
Step 12: Show the effect of the Gaussian Laplacian method (LoG).
Step 13: Show the result of the Canny method.
Step14: Show the result of the Zero Crossing method.

The first order exit is known as the Classical operator and Sobel operator, Prewitt operator, and Robert's Cross operator were the strategies for acquiring the Classical Edge. About sound images, these techniques are used to obtain effective edges. The benefits of using this Classical operator are high in performance and very easy to use.

4.1 Sobel Operator

Sobel operator is very simple in terms of smoothness and Gradient. A set of 3x3 convolution kernels and masks is used on the operator. Here the edges are enlarged using perpendicular edges. Two kernel processing methods are performed here one is allowed to rotate 90 degrees and the other one kernel stays in place. The Sobel edge detector process uses the mask shown in Figure. numerical measurement of the first outbound Bx and By and obtains edges using the Sobel proximity exit.

Bx

-1	-2	-1
0	0	0
1	2	1

By

-1	0	1
-2	0	2
-1	0	1

The standard syntax for the Sobel detector states

$$[g, t] = edge\ (c, \text{'Sobel'}, D, dir)$$

where c is the input image, D is the specified limit, and the dir specifies the preferred direction of the obtained edges. 'Horizontal', 'vertical' or 'both' (default).

4.2 Prewitt Operator

The Prewitt detector on the edge uses masks in Figure to do indirectly with the first output of Bx and By.

Bx

-1	-1	-1
0	0	0
1	1	1

By

-1	0	1
-1	0	1
-1	0	1

Its common syntax states:

$$[g,t] = edge(c,\text{'Prewitt'}, D, dir)$$

Obstacles to this activity are limited to Sobel limits. The Prewitt detector is a little easier to use electronically than the Sobel detector but relies on producing a certain level of noise effects.

4.3 Roberts Edge Detector
The Roberts edge detector uses a mask on the image. digital specification of the first output of Bx and By.

Bx

1	0
0	1

By

0	-1
1	0

Its common syntax says

$$[g,t] = edge(c,'Roberts',D,dir)$$

Obstacles to this function are similar to Sobel parameters. The Roberts Detector is one of the oldest digital image processing technologies.

4.4 Laplacian of Gaussian (LoG)
Laplacian is a 2-D isotropic measurement of the second placement of a second image. The image sets the rotating regions quickly so it is often used for edge detection. Laplacian is often useful for a photo that is initially smoothed out with a Gaussian Smoothing filter to minimize this particular sensitivity to sound. The user usually takes a gray level image as input and generates another gray image as output. The standard cost syntax for the LoG detector states

$$[g,t] = edge(c,'\log',D,sigma)$$

where sigma is a common deviation.

4.5 Canny Edge Detector
The canny operator is based on three criteria. Initially, the Gaussian work was used to smooth the image. The high value of the first acquisition is related to the low rate of the original acquisition. Changes in the grayscale are the result of a solid curve and a slight change with a weak end is accompanied by a second exit and a zero-crossing point. Therefore, these two thresholds are used to obtain strong edges and weak edges.

4.6 Zero Crossing
The Zero-Crossing operator uses the second output and contains the Laplacian operator. This is powerful in noisy data. This is the second direct result of a zero-resolution image-crossing project proposed by Haralick.
The standard syntax says:

$$[g,t] = edge(c,'zerocros\sin g',D,I)$$

where 'I' is a filter function.

5. Results and Discussion
We analyze different approaches to make comparisons between different edge detectors in images captured with different bright conditions. We took some grayscale photos in our smartphone and evaluated with preprocessing work such as reading the image, displaying the image, resizing the image, cropping the image, reshaping the edges of an image, applying different methods like Sobel, Prewitt, Roberts, canny, log and zero-crossing with different variations for that image as an experimental work using MATLAB application.

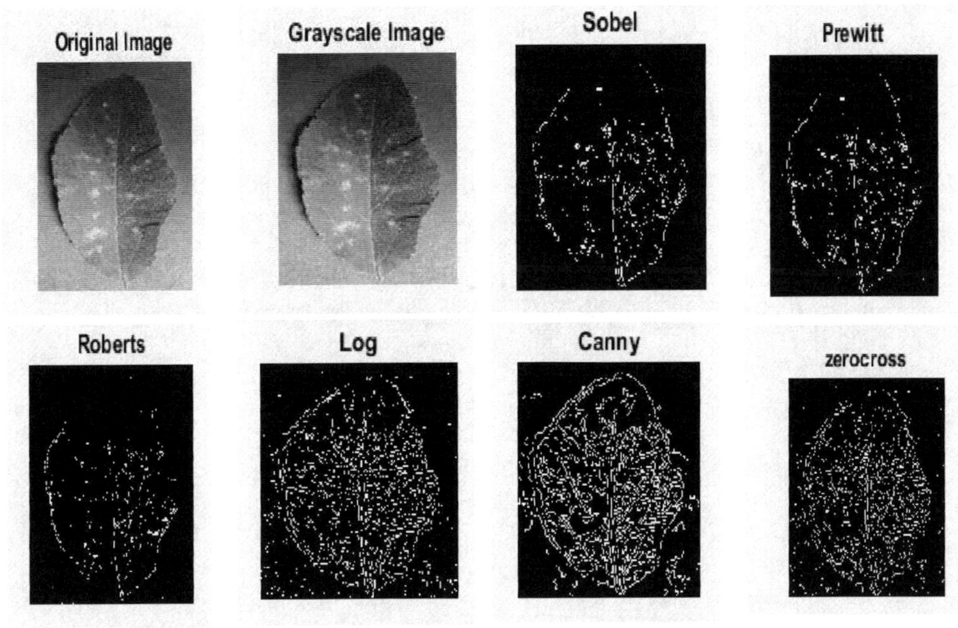

Figure 3. Comparison of various edge detection techniques

6. Conclusion

The marketplace for green technology is very in elevation as we are more and more aware of the environmental damage we cause. This awareness induces the market to introduce new and progressive green technologies. Altogether in everything, it is a natural success. New methods of green technology are increasingly being used and are getting better and more environmentally friendly. Therefore, the first step in the process of recognizing an object is the discovery of the edge. In this paper, comparisons are made with different edge detectors. In this regard, the tasks were performed using MATLAB. Operators processed in the First order are sensitive and while compared to Second-order exit operators these operators should not be better in operation. As a Canny operator that gives good results because it takes less time to calculate. More realistic edges were found in the lower comparative images compared to any other edge finders and for the actual application, this Canny operator can be used. Local performance problem solved by this Canny operator. Therefore, now related to this we conclude that the replaced operator has better results than any other operator.

7. References

[1] Mohd. Aquib Ansari, Diksha Kurchaniya and Manish Dixit,2017, *A Comprehensive Analysis of Image Edge Detection Techniques,* International Journal of Multimedia and Ubiquitous Engineering, Vol.12,No.11 (2017), ISSN: 1975-0080 IJMUE.

[2] Saket Bhardwaj and Ajay Mittal,2012, *A Survey on Various Edge Detector Techniques*, SciVerse ScienceDirect, Procedia Technology 4 (2012) 220 – 226.

[3] K. Bala Krishnan, Shiva Prakash Ranga, and Nageswara Guptha,2017, *A Survey on Different Edge Detection Techniques for Image Segmentation,* Indian Journal of Science and Technology, Vol 10(4), DOI: 10.17485/ijst/2017/v10i4/108963, ISSN (Print): 0974-6846, ISSN (Online): 0974-5645.

[4] Manzoor Ahmed Hashmani, Muhammad Umair, Syed Sajjad Hussain Rizvi, and Abdul Rehman Gilal,2020, *A Survey on Edge Detection based on recent Marine Horizon Line Detection Methods and their Applications,* 3rd International Conference on Computing, Mathematics and Engineering Technologies, 978-1-7281-4970-7/20/$31.00 ©2020 IEEE.

[5] Ramnarayan, Nikita Saklani, and Vasundhara Verma,2019, *A Review on Edge Detection Technique "Canny Edge Detection",* International Journal of Computer Applications (0975 – 8887) Volume 178 – No. 10.

[6] Diplaxmi R. Waghule and Dr. Rohini S. Ochawar,2014, *Overview on Edge Detection Methods,* International Conference on Electronic Systems, Signal Processing and Computing Technologies.

[7] Ghassan Mahmoud husien Amer and Dr. Ahmed Mohamed Abushaala,2015, *Edge Detection Methods*, 978-1-4799-8172-4/15/$31.00 ©2015 IEEE.

[8] Shubhashree Savant,2014, *A Review on Edge Detection Techniques for Image Segmentation*, International Journal of Computer Science and Information Technologies, Vol. 5 (4), 5898-5900.

[9] Muthukrishnan. R and M. Radha,2011, *Edge Detection Techniques for Image Segmentation*, International Journal of Computer Science & Information Technology (IJCSIT) Vol 3, No 6, 2011.3620.

[10] Khalid Alshalfan and Mohammed Zakariah,2020, *Edge Detection Enhancement Based on Filtering and Threshold Estimation*, International Journal of Engineering and Advanced Technology (IJEAT) ISSN: 2249 – 8958 (Online), Volume-9 Issue-5.

[11] Tasnuva Tasneem1, Zeenat Afroze2,2014, *Analysis of Edge Detection Technique by Varying Image Contrast*, IOSR Journal of Electrical and Electronics Engineering (IOSR-JEEE) e-ISSN: 2278-1676, p-ISSN: 2320-3331, Volume 9, Issue 5 Ver. II, PP 21-27 www.iosrjournals.org.

[12] Satbir Kaur and Ishpreet Singh,2016, *Comparison between Edge Detection Techniques*, International Journal of Computer Applications (0975 – 8887) Volume 145 – No.15.

[13] Tamanna Sahoo and Prof. Sandipan Pine,2016, *Design and Simulation of Various Edge Detection Techniques Using Matlab Simulink*, International conference on Signal Processing, Communication, Power and Embedded System (SCOPES).

[14] Muthukrishnan. R and M. Radha,2011, *Edge Detection Techniques for Image Segmentation*, International Journal of Computer Science & Information Technology (IJCSIT) Vol 3, No 6, 2011.3620.

[15] J. S. Owotogbe, T. S. Ibiyemi, and B. A. Adu,2019, *Edge Detection Techniques on Digital Images -A Review*, International Journal of Innovative Science and Research Technology, Volume 4, Issue 11, ISSN No: -2456-2165.

[16] Divyang Goswami,2015, *Edge Detection Technology using Image processing in Matlab*, International Journal on Recent and Innovation Trends in Computing and Communication, Volume: 3 Issue: 5, ISSN: 2321-8169, 3466 – 3471, Available @ http://www.ijritcc.org.

[17] Xin Bai, Mijia Yang, and Beena Ajmera,2020, *An Advanced Edge-Detection Method for Noncontact Structural Displacement Monitoring*, 20, 4941; doi:10.3390/s20174941, www.mdpi.com/journal/sensors.

[18] Xiaofeng Li, Hongshuang Jiao & Yanwei Wang,2020, *Edge detection algorithm of cancer image based on deep learning,* BIOENGINEERED, VOL. 11, NO. 1, 693–707, https://doi.org/10.1080/21655979.2020.1778913.

[19] Miroslav Hagara1, Radovan Stojanović, Tomáš Bagala, Peter Kubinec and Oldřich Ondráček,2020, *Grayscale image formats for edge detection and for its FPGA implementation*, Microprocessors and Microsystems, S0141-9331(19)30503-4, MICPRO 103056.

[20] Sunil Kumar, Amit Kumar Upadhyay, Preeti Dubey, and Sudeep Varshney,2021, *Comparative analysis for Edge Detection Techniques*, International Conference on Computing, Communication, and Intelligent Systems(ICCCIS), ISBN: 978-1-7281-8529-3/21/$31.00 ©2021 IEEE.

IOP Conf. Series: Earth and Environmental Science 1057 (2022) 012010 doi:10.1088/1755-1315/1057/1/012010

An experimental study of solar receiver under cloudy weather conditions

Sanjay N Havaldar[1], Kaustubh G Kulkarni[1], Nilesh T Dhokane[1] and Shrikant D Nimbalkar[1]

[1]School of Mech. Engg. Dr. Vishwanath Karad MIT World Peace University, 411038, Pune, India

Abstract. The performance of the central solar receiver should be properly examined for an effective design of a solar power or solar heating system. The performance of solar receiver is affected by three parameters. These are the amount of solar radiations, the state of the heating medium and the temperature of the surrounding air. Also, while evaluating the performance of a solar receiver, the ambient weather conditions should be taken into account. As a result, when a large amount of direct light strikes a solar receiver, the power generated increases as well. In contrast, as the amount of direct light falling on the solar receiver diminishes, so does the amount of electricity generated. As a result, it is generally understood that on sunny days, the performance of the central solar receiver will be higher, while on overcast days, the performance will be reduced. The solar receiver collects heat from diffused or reflecting light (when the sun is obscured by clouds) and absorbs reflective light from glossy or light-colored surfaces. The experimental results reveal that solar radiation collection performance is dependent on climatic circumstances; thus, when installing a solar receiver, it should be suitable for the specific region depending on weather conditions.

Keywords: Central Solar Receiver, Weather Conditions, Cloudy Days, Direct Light, Reflective Light

[1] **Correspondence email addresses: sanjay.havaldar@mitwpu.edu.in**

1. Introduction

The cost of fossil fuels and natural gas is rapidly rising, resulting in high power generation costs. Additionally, the use of fossil fuels around the world produces significant amounts of greenhouse gases, which are hazardous to all living organisms, flora and fauna, and the environment. As a result, an alternative power generating source that is more environmentally friendly, clean, and renewable is required [1]. In homes, industries, agriculture, transport, financial and educational sectors, energy (uninterrupted power) is a necessity for its daily operations [2]. Due to the wide applications of solar energy in the day-to-day life, it is important that solar energy can be accessible in all the weather conditions. The myth that solar energy can only be produced in direct sunlight is false. In cloudy days, solar receive solar scattered light and it's a challenge to design the receiver such it converts maximum solar energy in such situations.

The efficiency of a central solar receiver is dependent on the amount of solar radiation, the temperature of the surrounding air, and the state of the heating medium (inlet temperature and flow rate). It should be analyzed for optimal solar power system design. A solar power system should be used under the proper weather condition which has been likely to get succeeded in the use of such solar power systems for heating and power generation [3]. Solar energy harvesting is drawing the attention of a number of academics and academicians these days because it is a plentiful source of energy. Solar energy has also been utilized in countries such as Japan, China, and South Korea. Japan's energy self-sufficiency rate is as low as 7% without nuclear power [4]. Natural energy is being used for heating in Japan following the 2011 earthquake in the Tohoku region, which limited the use of electrical lines and fossil fuels. Despite advancements in home insulation, the energy created by heat is used more in northern regions. Solar energy is used for water heating in South Korea and China, and it can be used for a variety of purposes in other countries.

2. Literature review

Many of the research investigations are been piloted into the performance of the central solar receivers. In [5], a mathematical model of a flat plate solar collector was tested through a prototype experiment. A heating medium with varying temperature inputs was used to test the performance of a flat plate collector [6]. The amount of heat absorbed by the solar receiver was determined by the temperature of the surrounding air, the receiver's total heat loss, and the average temperature of the heating medium. The experiment was set up so that the heating medium's inlet temperature was maintained between 30 and 80 degrees Celsius at various times between 11:30 and 12:30 or 10:30 and 14:30, with a 1 second or 60 second time interval [7] examined the material transmittance, the rate of absorption and the convective heat conductivity of air, as well as the heat balance of the solar receiver's components. A flat-plate sun collector with a surface area of 2.72 m^2, an angle of 42 degrees, a height of 600 mm, and 50 mm of polyurethane insulation was used to examine the performance of a solar receiver linked to a hot-water tank in [8]. This setup can be used to compute the amount of solar energy gathered, as well as the temperature of the upper and lower parts of the tank, and the efficiency [9]. It shows a flat plate collector that can heat air, water, or a mixture of both.

A mathematical model for the heating modes of air and water was built to assess the impact of mass flow rates of 0.024 and 0.013 kg per second on thermal performance. The results of the experiment revealed that the mass flow rate is the most critical element affecting the solar receiver's efficiency, heat transfer, and outlet temperature. When the flow rate exceeds 0.10 kg per second, the heat removal factor rapidly increases. Testing was used to rate the performance of the thermo-syphon system in [10]. This performance evaluation should be based on the system efficiency test performed during the energy collection phase, which determines the energy absorption capacity in the energy collection phase. The system cooling loss test is performed during the cooling phase, which determines the rate of energy loss in the cooling phase. These tests were carried out in a natural setting. A formula for calculating the energy efficiency of a solar receiver has been devised with the help of the amount of solar energy gathered, efficiency, and heat loss.

This can be accomplished by employing a semi-empirical system efficiency model, performance testing, and rating, as well as developing a test system for the thermo- syphon system in the outdoors. In [11], five levels of sun irradiation and five levels of mass flow rate were used to test a solar collector with 16 U-type evacuated glass tubes. A solar simulator is used to calculate the performance of an evacuated solar receiver. The flow rate, the ambient temperature, and the amount of solar radiation are all altered by the solar simulator. A strategy for increasing the efficiency of solar radiation received by altering the flat plate collector's construction was discussed in [12]. The experiment was carried out to evaluate the efficiency and temperature of water, absorber plate, and glass-cover, as well as the solar receiver's thermal performance and heat balance. Furthermore, a model of transient heat transfer was used to evaluate the efficiency and temperature of water.

The flow rate was used to evaluate the performance of the ceramic collector in [13]. The experiment was carried out under typical operating conditions with two 140 liter hot water storage tanks, a circulating pump, and a microcontroller. Because of the higher efficiency and longer service life of the monolithic absorber, the ceramic collector can be used instead of standard solar panels. The typical full-sale metal solar collector working at ambient settings and its polymer-fabricated examples are discussed in [14]. A honeycomb multi-channel plate composed of polycarbonate was used to build the polymeric solar receiver. The results revealed that the efficiency of a polymeric collector was 8 to 15% lower than that of a traditional collector. A hybrid solar system was investigated in [15], and it was discovered that it could conserve more energy. This technique can be utilized for farrowing house heating. This study took place over the course of eight months. During this time, the heating demands for the warm season were met to a degree of around 70%, while the heating needs for the cold period were met to a degree of around 25% to 30%.

In [16], a U-pipe evacuated solar tube collector was tested, and its thermal performance was computed, and a novel model for heat transfer between the U-pipe and the fin was recommended. The correctness of this concept is determined by the solar collector's efficiency under various conditions. Based on the findings, a new solar collector design has been proposed. In [17], a model for predicting the dynamic thermal performance of a solar receiver was recommended [17]. Several studies for evaluating the performance of a solar receiver exist, but they are conducted under different experimental conditions or in different geographic regions, and they use different assessment approaches. As a result of the different results, comparing the performance of solar receivers is difficult. As a result, an experiment was carried out on solar receivers in the same geographical region, under the same experimental conditions in order to objectively compare the solar receivers.

[18] used the solar simulator to test the efficiency of a solar receiver under identical experimental conditions, and its validity was calculated using uncertainty analysis. While evaluating the effectiveness of the solar receiver under laboratory conditions, this research highlights the uncertainties generated by the measurement conditions. In it, the performance of a solar collector was examined based on geographical variables [19]. It was discovered that the thermal efficiency fluctuates with changes in wind speed, flow rate, radiation intensity, and so on. The average efficiency of radiation collection was computed based on fluctuations in local meteorological conditions throughout this experiment, which was carried out in ten Chinese towns. A mathematical model was investigated in [20] to determine differences in the efficiency of solar receivers dependent on wind speed, surrounding temperature, radiation intensity, surface area, and other factors, and to assess the likelihood of harnessing solar energy in cold weather regions. An experimental investigation into the applications of a central solar receiver under overcast conditions has been conducted. The performance of the solar receivers is being assessed in this study under sunny and cloudy conditions.

3. Methodology

The experimental setting has been set up in such a way that the amount of solar radiation, wind direction, wind speed, and surrounding temperature has all been calculated in order to test the central solar receiver's performance. In this experiment, two solar receivers were employed, each with an identical inlet temperature of the heating medium, which was controlled as the set-point of the heat exchanger's heating and cooling. Two heat exchangers are required to keep the inlet temperature of the heating medium constant. The heat transfer fluid was brine, and the receivers were made of copper tubing. Figure 1 shows the general experimental arrangement during the trials.

3.1. Measurement Conditions

The basic plate-type solar collector was employed in this investigation, and the experiment was run for a week with the same flow rate and inlet temperature of the heating medium to compare the performance of the two solar receivers. Each receiver was tested for a month by setting the input temperature to 40 and 60 °C, respectively, with a flow rate of 0.01 and 0.02 kg per second (see Table 1).

3.2. Solar receiver specifications

The flat plate collector, with dimensions of 2002 x 1002 x 60 mm, and area of 2.01 m^2, and a mass of roughly 37 kg, was employed in this investigation as the solar receiver. The experiment lasted about a month, with four sets of measurements being taken at one-week interval.

Table 1: Experimental measurement conditions

Measurement	Inlet Temperature (°C)	Flow Rate ($m^3 s^{-1}$)
1	40	0.01
2	40	0.01
3	60	0.02
4	60	0.02

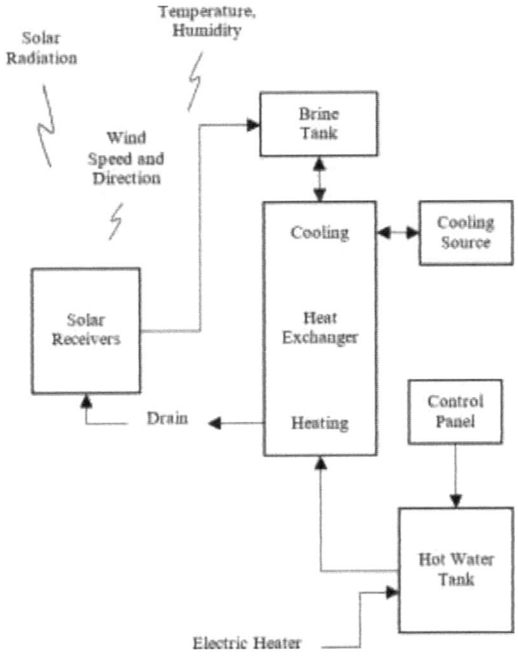

Figure 1: Layout of experimental arrangement.

3.3 Solar receiver specifications

The amount of solar radiation gathered by each solar receiver can be calculated using horizontal and inclined solar radiation as well as the heating medium's intake and output temperatures. In a nutshell, the difference between the intake and output temperatures, together with the flow rate, yields the amount of solar radiation collected, which may be stated mathematically as follows. watts (W) was calculated using Eqn. (1).

$$Q = C * \rho * m * (T_{out} - T_{in}) \quad (1)$$

where, C is the specific heat of heat transfer fluid in Joules per kg per degree Kelvin, ρ is the density of heat transfer fluid in kg per cubic meter, m is the mass flow rate in meters per second. The outlet temperature of (T_{out}) and the inlet temperature (T_{in}) of the heat transfer fluid is measured in degree centigrade.

3.4. *Solar collection efficiency*

The efficiency of solar collection by each of the solar receiver can be calculated with the horizontal and inclined solar radiation and the inlet and outlet temperature of the heating medium. In short, the ratio of the amount of solar radiation collection to the amount of the solar radiation gives the efficiency of solar collection which can be expressed mathematically in Eqn. (2).

$$\eta = \frac{Q}{I \times A} \quad (2)$$

where, n is the efficiency of solar collection, I is the solar radiation from the inclined surface in watts per square meters and A is the solar receiver area in square meter. The ratio of solar radiation collected to solar radiation received can be used to calculate the instantaneous efficiency of a solar receiver. The solar collection efficiency may was determined using the solar collection efficiency characteristic diagram which is given in Eqn. (3).

$$\eta = (b_0 - b_1) * (\Delta T / l) \quad (3)$$

In Eqn. (3), the difference in temperature (ΔT) is the difference between average temperature of heating fluid (T_w) and the air outside temperature (T_a). The average temperature of heating fluid is obtained by averaging inlet temperature (T_{in}) of the heating fluid and outlet temperature (T_{out})of the heating fluid, that is, $T_w = \frac{(T_{in}+T_{out})}{2}$. The solar absorbance is represented by b_0 and the collector heat-loss coefficient by b_1 in Eqn. (3) respectively and is measured in watts per square meters per Kelvin.

4. Discussion of results

The inlet temperature of the solar receiver is set to 60 °C and the flow rate is set to 0.02 cubic meters per second for both of the solar receiver. The outlet temperature of the second solar receiver is raised to 13 °C but the inlet temperature is same as that of 60 °C. The amount of solar radiation collection, the outlet temperature, the efficiency of the solar collection is compared for both the receivers which are represented graphically in Figure 2.

4.1 Results of the experiments

Figurers 2 to 5 are the time series plots of weather data, inlet and outlet temperatures of heat transfer fluid, solar radiation collection temperature and efficiency of the receivers in used the experiment respectively. From Figure 2, it can be found that solar radiation per square meter gradually increased from 8.00 hours and attained its peak by around 11.45 hours. It reduced gradually from its peak of 1 W/m^2 to zero at approximately 16 hours. During these hours (8.00 to 16.00 hours) of the day, the minimum outside temperature was approximately 15 °C to a maximum of 25 °C respectively (refer to Figure 2).

4.2 Solar receiver fluid temperature analysis

It is found that the efficiency of both the solar receivers remains approximately constant, as only a small amount of heat has been lost in the surrounding air. The maximum efficiency of the solar radiation collection of the second solar receiver $b_1 = 0.48$ which is less than that of the other solar receiver which is 0.70. Still, the solar radiation collection of the second solar receiver is better than the other receiver when the solar collection efficiency is high, like when the weather condition is cold or the solar radiation is less, since the collector heat-loss coefficient of the second solar receiver $b_0 = 0.57$, is less than the other collector which is 5.62. This has been graphically represented in Figure 3.

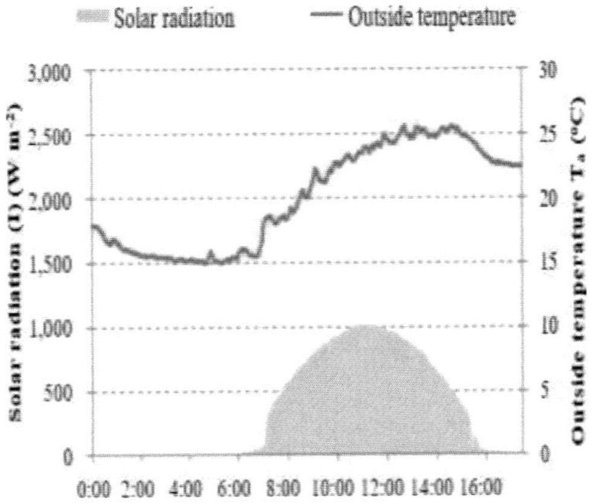

Figure 3. Time series of weather data recorded in the experiment.

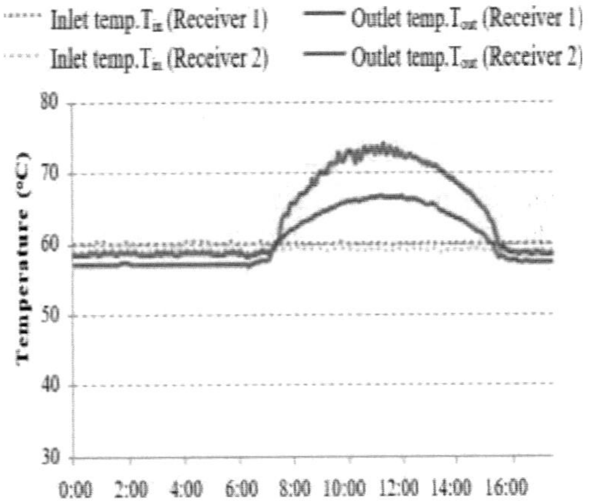

Figure 4. Time series of the inlet and outlet temperature of heat transfer fluid.

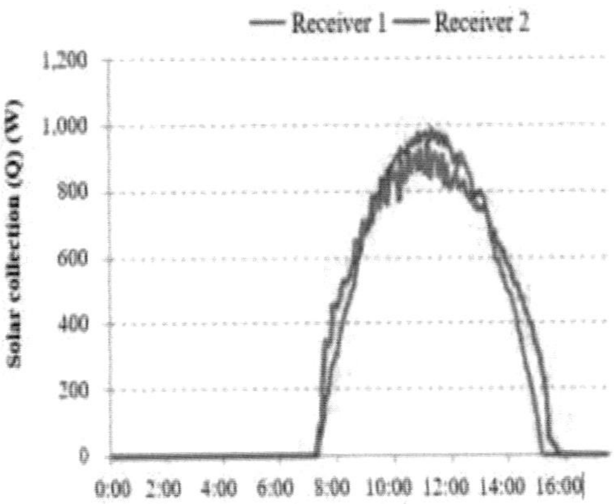

Figure 4. Time series of solar radiation collection on receivers in the experiment

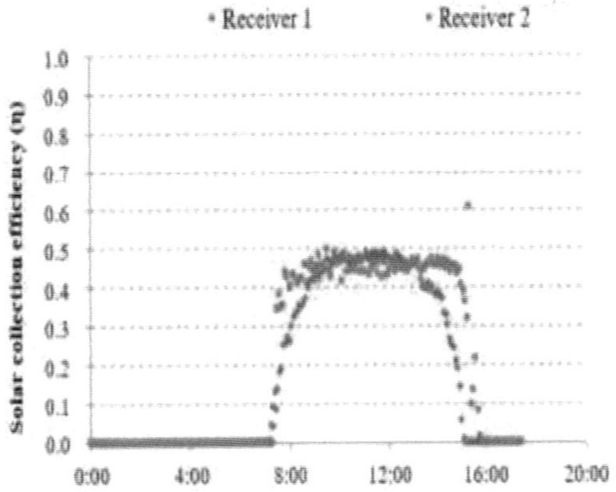

Figure 5. Time series of the solar collection efficiency for the receivers.

The comparison of both the solar receivers with inlet temperature 40 and 60 °C and the flow rate 0.01 and 0.02 m^3s^{-1} with respect to the solar absorbance and the collector heat loss coefficient is given in Table 2. From this result, it is found that, there is no substantial difference in the value of b_0 and b_1 when the inlet temperature has been changed. But when the collector is evacuated, then the slope increases but the efficiency of the solar collection decreases when compared with the flat plate collector which shows that the performance of the solar receiver depends on the weather condition.

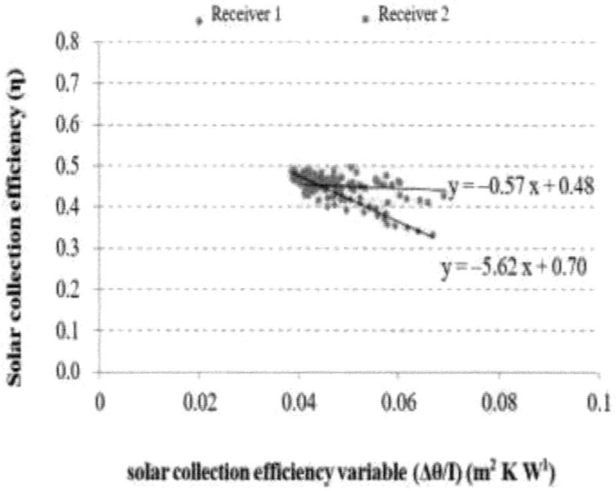

Figure 6: Comparison of the efficiency of the two solar collectors

Table 2: Amount of solar radiation collection per unit area

Solar Receiver	b_0	b_1	Root Mean Square Error: Inlet Temp. 40°C	Inlet Temp. 60 °C	Flow Rate 0.01 $m^3 s^{-1}$	Flow Rate 0.02 $m^3 s^{-1}$
1	0.59	-3.43	0.023	-	0.016	-
2	0.66	-4.06	0.018	0.017	-	0.009
Evacuated	0.46	-0.43	0.023	0.016	0.022	-

The solar radiation collected per unit area of the solar receiver has been calculated to examine the variation of solar collection efficiency based on the solar radiation and outside temperature. The solar receiver is fixed at 30 °C with an average temperature of the receiver of about 40 °C with direct solar radiation, which means no cloud or shadows near the solar receiver.

Table 3 shows the amount of solar radiation collection per unit area by the solar receivers under sunny and cloudy conditions. From this result, it found that the amounts of solar radiation collected in sunny days are more than that of the solar radiation collected in cloudy days. The result of the experimental analysis of amount of solar collection has been graphically represented in Figure 6.

Table 3: Amount of solar radiation collection per unit area

Solar Receiver	Weather collection	Amount of solar collection W/m^2
Receiver 1	Sunny	3.23
Receiver 2	Sunny	2.78
Receiver 1	Cloudy	2.56
Receiver 2	Cloudy	2.21

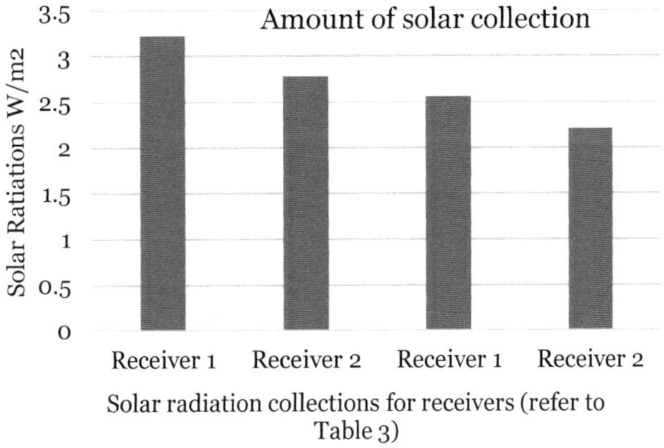

Figure 6: Comparison of amount of solar collection in different weather conditions

5. Conclusions

With the use of two solar receivers, this paper describes an experimental method for evaluating the performance of the central solar receiver system based on outside temperature, angle of inclination, and weather conditions. The efficiency of a solar receiver is determined in this experiment for varied intake and outlet temperatures of a heating medium, different flow rates, variable solar absorbance, and different collector heat-loss coefficients. In addition, the amount of solar radiation collected is computed and the results are supplied for various weather circumstances. The experimental results reveal that solar radiation collection performance is dependent on climatic circumstances; thus, when installing a solar receiver, it should be suitable for the specific region depending on weather conditions.

References

[1] Tripathi A K and Murthy C M 2017 *International Journal of Renewable Energy Research* **7** 2017–2017

[2] Kumar K E, Kumar S, Adelodun A A and Kim K H 2018 *Renew. Sustain. Energy Rev* **82** 894–900

[3] Wang R Z, Xy and Zy 2017 *Renew. Energy* **126** 1126–1140

[4] Matsumoto K and Shiraki H 2018 *Renew. Sustain. Energy Rev* **90** 391–401

[5] Zhou F, J, Ji, Cai J and Yu B 2017 *Appl. Therm. Eng* **118** 773–784

[6] Amer J K E H, Nayak G K and Sharma 1998 *Energy Convers. Manag* **39** 549–558

[7] Baba K and Kanayama 1989 *Sol. Wind Technol* **6** 51–58

[8] Michaelides P C I M and Eleftheriou 2011 *Exp. Therm. Fluid Sci* **35** 1002–1009

[9] Zhang D, Li, Gao, Wang and Nan 2016 *Appl. Therm. Eng* **108** 1126–1135

[10] Huang B J and Du S C 1991 *J. Sol. Energy Eng. Trans. ASME* **113** 172–172

[11] Nie X, Zhao S, Deng and Lin 2017 *Sol. Energy* **150** 192–201

[12] Wei L, Yuan D, Tang and Wu 2013 *Sol. Energy* **97** 19–25

[13] Zukowski M and Woroniak G 2017 *Sol. Energy* **146** 532–542

[14] Chen G, Doroshenko A, Koltun and Shestopalov 2015 *Sol. Energy* **115** 577–588

[15] Tamvakidis S, Firfiris A, Martzopoulou, Fragos V and Kotsopoulos T 2015 *Energy Build* **97** 162–174

[16] Gao Y, Fan and Zhang 2014 *Sol. Energy* **107** 714–727

[17] Deng J, Yang and Wang 2015 *Sol. Energy* **114** 418–426

[18] Sowmy D, Ara and Prado 2017 *Renew. Energy* **108** 644–651

[19] Fan M, Liang S, You, Zhang B, Yin and Wu 2018 *Appl. Energy* **221** 100–111

[20] Zheng W, Zhang S, You, Fu X and Zheng 2017 *Appl. Energy* **203** 938–947

IOP Conf. Series: Earth and Environmental Science 1057 (2022) 012011 doi:10.1088/1755-1315/1057/1/012011

Obesity: technology, exercise and diet

Roger Jaramillo[1], Alex Villao[2], Diana Merchán[3,4] and Jorge Jaramillo[5]

1 Universidad de Guayaquil, Guayaquil, Ecuador dr.rogerjaramillo_neuro89@hotmail.com, ORCID ID: 0000-0003-1245-2357
[2] Universidad de Guayaquil, Facultad de Ingeniería Química, Guayaquil, Ecuador alex.villaov@ug.edu.ec, https://orcid.org/ ORCID ID: 0000-0002-9143-2655
[3] Universidad de Guayaquil, Facultad de Ingeniería Química, Guayaquil, Ecuador diana.merchang@ug.edu.ec, https://orcid.org/ ORCID ID: 0000-0002-2352-3838
[4] Universidad de Valladolid, Escuela de Ingenierías Industriales, España,
[5] Universidad de Guayaquil, Facultad de Ingeniería Química, Guayaquil, Ecuador jorge.jaramilloor@ug.edu.ec, ORCID ID: 0000-0002-5754-5269

Abstract. Use of technology to accompany the treatment of obesity and exercise routine has become an ideal complement to carry historical data that are representative to measure the evolution of this disease. A bibliographic analysis of some studies was applied where it has successfully contributed to the use of APPs that are cornering the market because they not only measure the steps an individual takes, but also the calories consumed versus calories burned. APPs are a tool or support, but obesity is a complex disease that requires another type of support, not only an APP; Additionally, climate change affects food security, which causes the increase of this disease.

1. Introduction

New technologies provide important support in the treatment of certain diseases and complications that put people's health at risk. In recent years, both healthcare professionals and patients have seen ICTs as an inexhaustible source of solutions to the shortcomings of the different scenarios in the more conventional healthcare field (1). Mobile applications are an easily accessible and widely available tool. However, the most popular ones are not always the most reliable, useful and of high quality, nor do they always involve a health professional in their creation. Therefore, it is necessary to increase and/or publicize the existing regulation, publishing scientific recommendations or providing quality seals to the apps, which are known by users, and to promote the development of applications based on scientific evidence, ensuring the quality of the content, which provide useful tools. With all this, apps would be a beneficial, useful, safe, reliable and trustworthy tool for weight control and healthy diet (2).

Obesity is currently the most frequent chronic metabolic pathology, reaching epidemic proportions worldwide, and has become a major public health problem in both developed and less favored societies (3). Obesity occurs when the body contains too much fat, that is to say, there is a general hypertrophy of adipose tissue, a body mass index [BMI] in adults greater than 30 kg/m^2, according to the World Health Organization [WHO]. It is also considered a sign of obesity when the abdominal perimeter in men is greater than or equal to 102 cm and in women greater than or equal to 88 cm. Overweight and obesity are not the same thing. Obesity is the increase in body weight above 15% of the value considered normal, as the result of the increase body fat. Percentage of people who fit into these two categories, overweight and obesity, is determined by the Body Mass Index (calculation between the height and weight of the individual) and, in this sense, the WHO considers that there is overweight when the BMI is equal to or greater than 25 kg/m^2 (4).

IOP Conf. Series: Earth and Environmental Science 1057 (2022) 012011 doi:10.1088/1755-1315/1057/1/012011

Climate impacts such as extreme weather events, the spread of pests and diseases, biodiversity loss, ecosystem degradation and water scarcity will worsen as the planet warms. These impacts can be detrimental to food security and livelihoods and lead to forced migrations (5). Long-term climate change, particularly global warming, could affect agriculture in a variety of ways, almost all of which pose a risk to the food security of the world's most vulnerable people (6). In a state of food insecurity, the population resorts to junk food, which has a high energy content and in complex interaction with other factors contributes to the prevalence of obesity in the pediatric age group (7). Biodiversity for food and agriculture is indispensable for food security, sustainable development, and the provision of many vital ecosystem services. Biodiversity makes production systems and livelihoods more resilient to shocks and adverse factors, including the effects of climate change (8).

As a result of the era of information technology, personal technologies, and the ease of data management, it is inevitable to associate advances in the field of health with advances in the quality of life of mankind, since the health sector is one of those that have benefited the most throughout history from these advances. One of the objectives of any health system is to try to improve the condition of its population, and in the case of pathologies, to provide useful tools that meet the healthcare requirements. Health systems are successful when they manage to provide access to people in need of health services according to their needs, ensuring that they are satisfied with the services provided, making sure that these really improve their quality of life (9)

Use of ICTs reaches all areas of the school world, including the subject of Physical Education. But working on it remotely and without going to school can seem a bit complicated. However, this selection of applications is ideal for taking sports education home. They are available on different platforms and encourage the practice of sport and a healthy lifestyle (10)

2. Methodology

Bibliographic methodology was applied for the review and analysis of research on the integration of technology, exercise and hypocaloric diet in the treatment of obesity. Figure 1 shows the search equations used for the research case.

Table 1. Search equations used for the research case

Ecuación de búsqueda	Resultado
Obesity AND diet	457
Technology AND diet hypocaloric	47
Obesity AND exercise	97
Obesity AND Technology	77
Obesity AND Climate Change	85

3. Results

Foot pressure technology has been utilized to study the influence of over-weight and obesity on the ankle joint which because of gaining weight lead to change gait characteristics, particularly foot ailments, and has been used in academic teaching in the Medical and Allied Health fields for over 30 years. However, earlier sensors often required an indoor setup as they were not portable. Advanced technology provides the means to take the system outdoors, thus the having the ability to determine the impact of walking-paths terrain on foot pressure and on the ankle joint. Over-weight and obesity cause a number of adverse walking issues for the ankle as extra weight is loaded onto this vital joint component. In addition, footpath terrain may be uneven and thus plays a role in the characteristics of gait and subsequently has an impact on the ankle (11).

Balanceat app was created with the main objective of helping and serving as a tool for the general population in the acquisition of healthy eating habits based on a balanced diet, using traditional recipes and products that can be found in the supermarket and also includes nutritional labeling. These characteristics make Balanceat a useful and endorsed tool available to all consumers. Another of the objectives of this app is that it can be used as a reference tool for nutritional support and advice for health professionals (12).

In the specific case of digital health apps, the data show that not only are they here to stay, but that their use has skyrocketed in recent times. One need only take a look at the iTunes Store or Play Store to see the wide range of apps available. According to the report "The Growing Value of Digital Health", prepared by the IQVIA Institute of Human Information Sciences, it is estimated that the number of these health apps worldwide exceeds 318,000 (almost double the number available two years ago), with some 200 more being added every day. Although the majority of the subject matter of the apps developed in the health field is related to general wellbeing, a "subgroup" is emerging consisting of those related to health status monitoring, often associated with patient care and currently representing 40% of all apps in the health field (13) .

Sedentary lifestyles and poor nutrition have gained ground since the population gained access to cars, technology and fast food, also known as junk food, which is characterised by its high fat and carbohydrate content. In recent decades, bariatric surgery has been used worldwide as a treatment for obesity, because obesity has spread as an epidemic in the developed world, this disease is determined as a chronic condition, and therefore, it is not possible to cure it, it can only be controlled, but never disappear (14)

A study of weight management PPPs found that only 0.8% (3/393) of the applications had undergone scientific evaluation and 0.3% (1/393) included the participation of health care experts. The paper concludes that commercial mobile weight loss/management apps lack evidence-based features, do not include health care experts in their development process, and have not undergone rigorous scientific testing. This calls into question the validity of the apps' claims of efficacy and safety. Collaboration between developers, researchers, clinicians, and patients to develop and test high-quality, evidence-based mobile apps for weight loss would be desirable (15). Several researchers have expressed concern about the direct and proportional relationship between climate change, food insecurity and obesity. Diabetes and obesity are considered as main diseases that have appeared with the changes in the diets of different generations (16).

Environmental consequences of food systems are beginning to mark their place in public health agendas. The definition of sustainability in food and nutrition is the characteristic or state with which the food needs of the current population or of a local population can be met without compromising the ability of future generations or populations in other regions to meet their own needs, and aquaculture is not immune to the effects of climate change; access to fish and seafood is asymmetrical among the different social strata due to their high prices (7). The abuse of high-calorie and high-protein diets causes millions of consumers in rich countries to die of obesity-related diseases (cardiovascular, cerebrovascular, diabetes and cancer), which the European Union (EU), Latin America and the

Caribbean (LAC) consider to be a global pandemic. Worldwide, some 41 million children under 5 years of age are overweight or obese (17). Irresponsible human activity leads to climate change, which, together with genetic factors, sedentary lifestyles, and other environmental factors, has a bidirectional impact on childhood obesity. On the one hand, it causes food insecurity due to the increase in the price of healthy food and the increase in the consumption of junk food, and on the other hand, the condition of being obese makes children more vulnerable to the effects of climate change because they are the ones who have the greatest difficulty in adapting (7).

The increase in childhood obesity associated with sedentary behaviours, increased by the use new technologies and a deficient diet, generates unparalleled levels of overweight and obesity. Therefore, the practice of physical activity should be understood as one of the forms of primary and secondary prevention for the reduction of chronic diseases and premature death (18), the use of technology could play against, due to the sedentary lifestyle it causes or promote anorexia because there are APPs that count the calories of each food.

Adolescents with disabilities have complex support needs and priorities around weight management and healthy lifestyles. Current PPPs have not considered within their functionalities the variety of disabilities. (19), this and other factors are important and necessary to validate a PPP that meets the needs to be accompanied within a process to treat obesity. A study from Case Western Reserve University (USA) links exposure to air pollution with an increased risk of overweight. It indicates that environmental factors may increase the risk of developing obesity and diabetes. (20). Overweight and obesity are chronic multifactorial diseases in which genetic and environmental factors are involved, manifested by pathological expansion of body adipose deposits(1) and with an abnormal or excessive accumulation of fat, characterized by a body mass index (BMI) greater than 25 kg/m2 (overweight) or ≥30 kg/m2 (obesity) in adults (21).

Cost of the APPs is given by basic and premium plans that fluctuate between \$10.00 and \$70.00 the cost effectiveness [CE] of these applications is given by the number of pounds that the person loses per month, for example two people acquire the premium plan at \$70.00 each, the first person loses only 18 pounds in the month, and the second person loses 12 pounds in the month, then the CE would be:

$$CE = \frac{V}{IE}$$

CE, Cost effectiveness; V. Value of APP ; IE, Performance indicator (lb)

$CE1 = \frac{70}{18} = 3.89$, the cost effectiveness in the first case is \$3.89

$CE1 = \frac{70}{12} = 5.83$, the cost effectiveness in the first case is \$5.83

The less weight you lose, the more expensive the application becomes, whoever has the greatest impact in achieving the objective will get a lower CE. Path towards a safe and environmentally friendly agrifood system is to modify the way we produce, distribute and use food, to use this food for human consumption, to implement production systems that do not waste so much food, and to modify our diet. Avoiding excess meat consumption, which implies reducing its consumption in some sectors of the population, is key to taking care of the planet and our health (22).

4. Conclusion

Accelerated growth of the development of APPs as a technological tool to support or accompany the treatment of obesity through the control of exercise and diet routines has a commercial approach, these applications do not have the endorsement of medical experts to guarantee the functionality they offer, however, market acceptance is high because losing weight has become a business strategy for some fields. Climate change limits access to nutritious food and has decreased biodiversity, which is a factor in the increase in obesity worldwide.

References

1. ¿Son eficaces y seguras las apps para tratar la obesidad? [Internet]. 2017 abr. Disponible en: https://www.consalud.es/saludigital/65/son-eficaces-y-seguras-las-apps-para-tratar-la-obesidad_41534_102.html

2. Dolado Martín C, Berlanga Fernández S, Salcedo M. Aplicaciones móviles de control de peso y dieta saludable, ¿todas valen? 2018; Disponible en: https://www.enfermeria21.com/revistas/ridec/articulo/27150/aplicaciones-moviles-de-control-de-peso-y-dieta-saludable-todas-valen/

3. DURAN POVEDA MGG, JUAN; RUI. Principios de la obesidad mórbida [Internet]. Place of publication not identified: DYKINSON.; 2017 [citado 20 de febrero de 2022]. Disponible en: http://www.digitaliapublishing.com/a/47831/

4. Díaz Rico JJ. El libro negro de los secretos de la obesidad. [Internet]. Madrid: Bubok Publishing S.L.; 2018 [citado 20 de febrero de 2022]. Disponible en: https://public.ebookcentral.proquest.com/choice/publicfullrecord.aspx?p=5425578

5. FAO. El trabajo de la FAO sobre el cambio climático: Conferencia de las Naciones Unidas sobre el cambio climático 2019 [Internet]. Rome, Italy: FAO; 2019 [citado 20 de febrero de 2022]. 40 p. Disponible en: https://www.fao.org/publications/card/es/c/CA7126ES/

6. La agricultura y los cambios climáticos: la función de la FAO [Internet]. [citado 19 de febrero de 2022]. Disponible en: https://www.fao.org/noticias/1997/971201-s.htm

7. Piña Borrego CE. Cambio climático, inseguridad alimentaria y obesidad infantil. Rev Cuba Salud Pública [Internet]. 13 de enero de 2020 [citado 19 de febrero de 2022];45:e1964. Disponible en: https://www.scielosp.org/article/rcsp/2019.v45n3/e1964/

8. FAO. El estado de la biodiversidad para la alimentación y la agricultura en el mundo - resumen [Internet]. Rome, Italy: FAO; 2019 [citado 20 de febrero de 2022]. 16 p. Disponible en: https://www.fao.org/publications/card/es/c/CA3229ES/

9. Álvarez F. Promoción del control del sobrepeso y la obesidad mediante el uso de tecnología móvil en la asistencia sanitaria [Internet]. 2020 [citado 21 de febrero de 2022]. Disponible en: https://academica-e.unavarra.es/bitstream/handle/2454/37809/ALVAREZ_131417_IMPRESO%20TFM.pdf?sequence=1&isAllowed=y

10. Mejores aplicaciones de Educación Física | EDUCACIÓN 3.0 [Internet]. 2020 [citado 21 de febrero de 2022]. Disponible en: https://www.educaciontrespuntocero.com/recursos/aplicaciones-de-educacion-fisica/

11. Al-Magsoosi SKR, Chong AK. Advanced Foot Pressure Technology for Demonstrating the Influence of Over-weight and Obesity on Ankle Joint. En: Proceedings of the 2019 2nd International Conference on Information Science and Systems [Internet]. New York, NY, USA: Association for Computing Machinery; 2019 [citado 20 de febrero de 2022]. p. 195-9. (ICISS 2019). Disponible en: https://doi.org/10.1145/3322645.3322661

12. Hernández EM. Impacto de las APPs de nutrición en la mejora de los hábitos alimentarios. :39. Disponible en: http://hdl.handle.net/11441/65245

13. Ventajas y asignaturas pendientes de las apps en el sector salud [Internet]. [citado 21 de febrero de 2022]. Disponible en: https://www.elsevier.com/es-es/connect/ehealth/ventajas-y-asignaturas-pendientes-de-las-apps-en-el-sector-salud

14. Astete Arriagada G. Cirugía bariátrica, cirugía de la obesidad: No es solo bajar de peso, es un cambio en el estilo de vida [Internet]. 2018 [citado 21 de marzo de 2022]. Disponible en: https://elibro.net/es/ereader/elibrodemo/196178

15. Rivera J, McPherson A, Hamilton J, Birken C, Coons M, Iyer S, et al. Mobile Apps for Weight Management: A Scoping Review. JMIR MHealth UHealth [Internet]. 26 de julio de 2016 [citado 22 de marzo de 2022];4(3):e5115. Disponible en: https://mhealth.jmir.org/2016/3/e87

16. Apps para la pérdida de peso ¿que hay en el mercado? [Internet]. [citado 31 de enero de 2022]. Disponible en: https://www.fundacionisys.org/es/blogs/mhealth/164-apps-para-la-perdida-de-peso-que-hay-en-el-mercado

17. Report of the commission on ending childhood obesity [Internet]. [citado 20 de febrero de 2022]. Disponible en: https://www.who.int/publications-detail-redirect/9789241510066

18. Bellanger A-P, Reboux G, Murat J-B, Bex V, Millon L. Detection of Aspergillus fumigatus by quantitative polymerase chain reaction in air samples impacted on low-melt agar. Am J Infect Control. septiembre de 2020;38(3):195-8.

19. Oake M, Stinson J. "Don't sweat it buddy, it's OK": an exploration of the needs of adolescents with disabilities when designing a mobile application for weight management and healthy lifestyles. Taylor Francis Ltd [Internet]. mayo de 2020; Disponible en: https://bv.unir.net:2144/doi/full/10.1080/09638288.2018.1530804

20. Diabetes F para la. https://www.fundaciondiabetes.org/general/noticia/15011/la-contaminacion-aumenta-el-riesgo-de-desarrollar-obesidad-y-diabetes [Internet]. [citado 31 de enero de 2022]. Disponible en: https://www.fundaciondiabetes.org/general/noticia/15011/la-contaminacion-aumenta-el-riesgo-de-desarrollar-obesidad-y-diabetes

21. Gutiérrez Valverde JM, Guevara Valtier MC, Enríquez Reyna MC, Paz Morales M de los Á, Hernández del Ángel MA, Landeros Olvera EA. Estudio en familias: factores ambientales y culturales asociados al sobrepeso y obesidad. Enferm Glob [Internet]. 28 de diciembre de 2016 [citado 31 de enero de 2022];16(1):1. Disponible en: http://revistas.um.es/eglobal/article/view/253811

22. Garibaldi LA, Andersson G, Ferrari CF, Pérez-Méndez N. Seguridad alimentaria, medio ambiente y nuestros hábitos de consumo. Ecol Austral [Internet]. 16 de noviembre de 2018 [citado 20 de febrero de 2022];28(3):572-80. Disponible en: http://ojs.ecologiaaustral.com.ar/index.php/Ecologia_Austral/article/view/768

A DISCUSSION ON THE IMPACTS OF ECOLOGICAL INTELLECTUAL CAPITAL IN GREEN MANAGEMENT

M.Suba[1], S.B. Inayath Ahamed[2]
[1]School of Management, Kalasalingam Academy of Research and Education, Krishnankoil.
suba.gajalakshmi@gmail.com
[2]School of Management, Kalasalingam Academy of Research and Education, Krishnankoil.
Inayathahamed@gmail.com
Corresponding Author: [1*] suba.gajalakshmi@gmail.com

Abstract. The ecological manpower administration is a process that also fully covered the Manpower administration procedures. eco-friendly manpower management is a major part of the profit creation and shortening the wastages in day-to-day business routines, as that concerns the scholar life to examine the relationships between the Green management and the financial movement of a firm. Most of the studies have some particular sets of factors that determine the functions of green human resource management. This study adds some specifications related to the financial outcomes by using or following ecological management practices. This is a conceptual study of the previous year's literature.

Keywords: Environmental manpower management, Green Management, Intellectual capital.

1. INTRODUCTION

In the present financial condition, incorporeal assets become a significant consideration for the productive advantages of the organization. The market estimation of the organization is commensurate with the commercial resources and intellectual capital (Johnson, 1999) (16). Whatever, the period of the understanding economy, organizational intellectual capital is every time higher than economical capital. Addedly, the population of network and duties sectors, the difference among market estimation, and the originality of the company have been rising regularly; sequent, original estimation of the organization is not exact on its economic reports. Seeing that financial methods cannot show the correct market values of the organization at present, the valuation changes from the prior valuation of corporeal assets to the reaction of corporeal assets. Accordingly, present studies with heed to intellectual capital have drawn much exploration currently on evaluating the original estimation of the organization the expression of intangible assets. The current studies concerning intellectual capital have drawn much understanding at present on how to estimate the original value of firms (Stewart, 1994) (45).

2. GREEN HUMAN RESOURCE MANAGEMENT

Researchers connect ecological manpower management to supportable and ecological administration utilizing the terms ecological human asset administration feasible manpower administration and ecological human asset administration to allude to a similar dimension of human asset the board (Renwick, D.W., Redman, T., & Maguire, S. (2013) (40). Masri and Jaaron (2017: 29) characterize ecological manpower administrations as "utilizing Human resource strategies to support ecological economical practices and increment representative's responsibility to the concern of natural maintainability." Consequently, Renwick et al., (2013) (40) eco-friendly human resource management incorporates receiving climate well-disposed HR activities; which may remember the part of HRM for contamination counteraction root of the association's working cycles. Mampra (2013: 26) characterizes Ecological manpower administration as "the utilization of HRM arrangements to support the reasonable utilization of assets inside business ventures and advance the reason for environmentalism which further boots up representative spirit and fulfillment."

This implies to the workouts' ecological techniques, associations ought to be furnished with eco-friendly representatives (Ooi et al., 2017) (33). Jirawuttinunt (2018: 21) expresses that ecological manpower administrations are "the utilization of ecological manpower management strategies to help the supportable utilization of assets inside the association and leads ecological administration benefits." Mathapati (2013) (30) presents the word ecological worker as the result of eco-friendly manpower management. He expresses that ecological manpower administration is straightforwardly liable for making a green worker who comprehends, appreciates, and rehearses green activities and keeps their ecological aims by the way of Human resource management interaction of selecting, engaging, preparing, remaking, creating, and driving the organization's human resources and business. Opatha and Arulrajah (2014) (34) characterize Ecological manpower administration as strategies, standards, and frameworks that build workers of the association eco-friendly substructure the particulars, consumables, common habitat, and the firm. The momentum analysis utilizes the word Ecological management to allude to utilizing the human resource structure to furnish the association with a green climate situated labor force that comprehends, appreciates, and rehearses green activity and keeps up its natural destinations all through the human force strategies to lead feasible administration corollaries.

3. OBJECTIVES OF GREEN HUMAN RESOURCE MANAGEMENT

The focal point of ecological manpower administration is to understand the financial benefits by creating the workers responsive to the particulars of ecological administration that is the basic actions of green human resource management, and its routines and how does it make affect the ecology. The employ originally stimulates the workers and improves knowledge of concert in them for creature a role of the green progress.

4. CONCEPTUAL MODEL

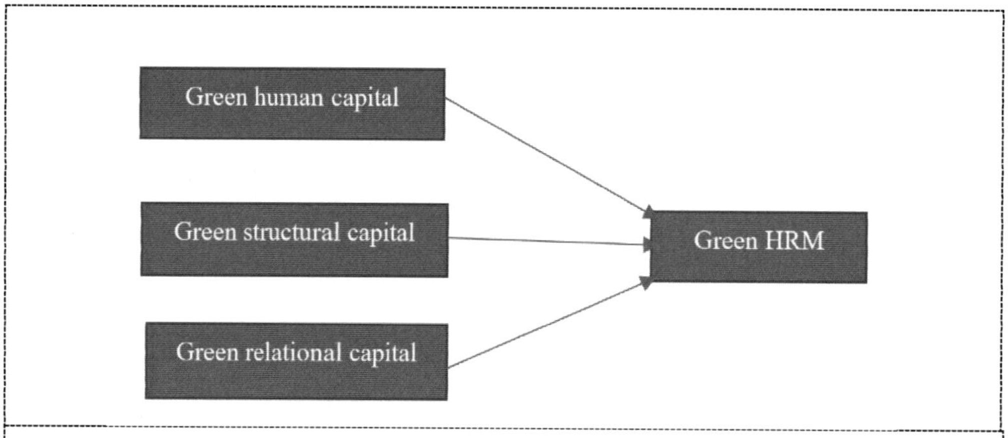

Figure 1. illustrates a proposed model for cold chain logistics in a organizational settings

5. HIGHLIGHTS OF GREEN MANAGEMENT

Ecological human power administration performing a considerable part in fields to improve the eco-connected concerns. Industries should reschedule workforce policies and applies to raise knowledge around the ecology, and implement values connected to ecological safety. Green human resource management can similarly assist the workers; producers in structure produce product image and goodwill.

Companies required organizing an ecological assessment, thus moving the commercial values, and ideas on waste it will also create workers and the public awake on the usage of global reserves most sparingly and committed to ecological yields.

1. Assisting organizations to lower the expenses deprived of dropping their aptitudes.
2. Industries have various development choices by practicing green and making a fresh ecological rule which supports massive functional reserves by lessening their toxin emission.
3. The green practices aiming complex worker work satisfaction and engagement which directs to huge efficiency and durability.
4. Make a value of taking care of the happiness and strength of associate employees.
5. Development in the retaining degree of the worker
6. Development of society image. Anytime a company includes green creativity in its workstation, it can utilize the moment to make good society connectivity. Companies can improve ecological offerings to the mess media over press proclamations to gain the attraction of latent consumers and probable new deals.
7. Improves workers' confidence.
8. Development in appealing to good workers. Dolan's (1997) research of USA MBA students revealed that more of the worker's earrings less amount to performing for ecologically accountable companies.
9. Decreasing the eco-friendly impression of the organization.
10. Developed affordability and raising through all performance.
11. Lowering the usage amount importantly: Small scale organizations can considerably lessen the usage expenses by utilizing skills that are power- proficient and reduce wastages.
12. Refunds and tax paybacks: Practicing green is simpler with the help of governments, local municipalities, water supply experts, and electronics fields that gives tax enticement and refunds.
13. Raising commercial choice: many government offices, money-making commercials, and non-money-making organizations that encounter particular green stands can offer on their agreements, some also decree that their buying departments only purchase eco-friendly products otherwise use products and amenities sold by institutions that see certain green canons.
14. Reducing environmental impairment: cheering workers, by guiding and recompense to discover methods to lessen ecological descriptive resources.
15. Nowadays most studied and prosperous buyers watch for the associations that espouse ec0-friendly values. Companies following eco-friendly workforce practices too massively promote.
16. Following the green concept in receiving eco-friendly practiser goods and eco-friendly saving and from wastage reduction. The improvement of some values can also flowingly promote customer fulfilment.

6. COMMON FACTORS OF THE GREEN STUDY

Researchers have introduced diverse ecological manpower management rehearses. For instance, Milliman and Clair (1996) need four ventures for a natural human resource administration model: (1) give ecological aims as a lead; (2) treat workers to share their natural objectives; (3) assess worker ecological execution; and (4) perceive representative natural exercises utilizing reward programs. Every day and Huang (2001) recommends the accompanying four-measurement reasonable structure for carrying out components of a human asset in the natural administration framework: help of experienced, preparing, strengthening, then compensations as principal segments of ecological HR. Renwick et al. (2013) (40) report that green recruiting (in which associations are intending to prepare themselves besides representatives over explicit natural abilities and surrounded by conventional affectability en route to the climate); environmental preparation over inclusion (while associations are quick to create ecological capabilities what's more, abilities and connect with their representatives in green practices); and green execution the board and pay (in which associations evaluate their worker execution utilizing, as one of the fundamental models, their green practices and remunerating those practices) are decidedly identified with the association's ecological exhibition

They further propose that (1) choosing; (2) enrolling; (3) preparing; and (4) creating ecological information can be considered as the segments of ecological management. Numerous researchers think about enrollment; execution of the executives and evaluation; preparation and advancement; work relations; and pay and award approaches as the amazing assets for adjusting workers to an association's natural procedure (Sharma, K. (2016) (43); Nawaratne, N.N.J., et al., 2015; Tiwari, P. et al., 2015;) (3) Jia, J. et al., (2017). The reason as most examinations alludes to the accompanying GHRM rehearses (1) enrollment and choice, (2) preparing, (3) execution of the board, (4) reward and award frameworks, and (5) association. In this exploration, GHRM is seen as made out of four practices are ecological enlistment and determination; eco-friendly preparation and enhancement; ecofriendly execution of the executives; ecological remuneration and go with the award, and it appears to be these conventions are the major ordinarily utilized and referenced eco-friendly manpower management rehearses in writing.

7. MONETARY MAINTAINABILITY

Monetary maintainability is hard to characterize and surprisingly more mind-boggling to gauge. Stringently talking, for an association to be considered monetarily feasible it should have adequate monetary assets to include its present commitments allied with functional expenses as well as obligations, combined with collect lasting-haul venture combined with improvement proposals (SaDhan, 2005). Be that as it may, the description might fluctuate generally between for-benefit and non-benefit associations. León, P. (2001; 25) characterizes monetary supportability on the point of "the ability to get incomes because of an interest, to maintain profitable cycles at a consistent or developing rate to deliver results and to acquire an excess." This relates to accomplishing long-haul monetary supportability through productivity and income excess. Emmanuel (2015) (12) characterizes monetary supportability as the capacity of a venture, or an association to keep up more extensive wellsprings of subsidizing to give standard administrations to its customers over the long haul. (Stilmant, D. et al., 2013)(24) recommend with the accomplishment of monetary maintainability in the organizational benefits the accomplishment of an ideal monetary exhibition or potentially ideal monetary position, which in turn, ought to take into consideration to stay in the presence (activity) for a long time to come. In like manner, monetary supportability requires keeping up subsidizing streams to guarantee the progression of the operational interaction, resource supportability (Almqvist, R. et al., 2014) (48), and dissolvability (Ayuba, B. et al., 2015) (17). Benefit keeps on being the most significant pointer of monetary manageability as the endurance of any business depends, generally, on its occasional productivity (Alshehhi et al., 2018 (1); Umobong, A. (2015) (47). Henceforth, contemplates have adjusted monetary manageability to long haul benefit (Isibor, A.A. et al., 2017 (32); Umobong, A. (2015) (47). Dominguez, K.M. et al., 2012;). Different pointers on monetary maintainability incorporate resource supportability; proficiency; liquidity; dissolvability; dependability; and development. (Alshehhi et al., 2018;(1) Umobong, 2015) (47).

8. INTELLECTUAL CAPITAL

The ecological intellectual capital is the capital that is compassed to the opulence of the company. It adds relational capital, manpower capital, and procedural capital. Everything or its meaning is compasses productive benefits and will be utilized for making currency that can be included in the intellectual capital of the working entity. Intellectual capital is the amount of all information; suggesting that information previous in various stages of interior or exterior the association should be considered on scholarly investment as resources. And the theoretical ideas on numerous information items and cycles, in blend with the expanding significance of their worth incorporate monetary records, prompt a developing revenue in the administration of scholarly capital. Making, molding, and refreshing the supply of scholarly investment needs the plan of an essential aim, what mixes each of the three elements of scholarly investments (Human, Structural and Relational Capital) inside the authoritative setting by the way of investigation as well as abuse, estimation, revealing. consequently, the hierarchical worth in scholarly capital is created through a continuous and rising interaction zeroed in on the ability to use, create and convert the determinants. Administration of scholarly investment is conceptualized as happening using various stage measures, administered by a transformative rationale. The scholarly capital administration is characterized as a pattern of 4 related arrangements of functions: action plan arrangements, peregrination, enslavement, magnitude, as well as recounting of initial resources.

Scholarly resource assumes a significant part in creating an incentive for organizations, just as for the worldwide economy. With the comprehension of the worth of the scholarly capital, organizations have grown better approaches to overseeing existing information. This had had numerous essential ramifications for associations' business tasks. The essential job of the scholarly resources is in this manner continually developing and associations are searching for more successful scholarly capital administration rehearses. On account of the significance of scholarly capital, these days we are discussing information laborers who are continually altering, using, and making new information to enhance the tasks of organizations. Therefore, it tends to be an administration of scholarly capital that is a critical determinant in associations' esteem formation. Here and there scholarly resource is additionally connected with the term information the executives. Explores have attempted to discover the association between these terms.

Scholarly capital is the all-out supply of all the immaterial resources, information, and abilities of an organization that could make esteems or cutthroat benefits, to accomplish its magnificent objectives (Masoulas, 1998) (28). Furthermore, educated capital was likewise characterized as the absolute supplies of every elusive resource, what's more, abilities in an organization that can make esteems or on the other hand upper hands (M. S. Malone: et al., 1997 (11); Stewart, T. A.: 1994) (45). (Stewart, T. A.: 1994) (45) a characterized intellectual resource is an all-out supply of the system information, data, advancements, scholarly effects right, insight, association grasping as well as ability, group correspondence frameworks, customer relations, and brands that can make value to the business association. There were two kinds of intellectual capital examinations: scholarly resources and scholarly resource estimation. The learns about scholarly capital administration zeroed in on the most proficient method to oversee immaterial resources, information stocks, and capacities of an organization to make esteem or upper hands. Nonetheless, the investigations of scholarly capital estimation focused on a social occasion, assembling, dissecting, and evaluating non-monetary data to quantify immaterial resources of firms that are inadequate on budget reports. Consequently, the revelation in scholarly resources might be thought of as beneficial data of budget summaries of an organization. Hence, the idea of scholarly capital can carry an upset to the conventional bookkeeping framework (M. S. Malone: et al., 1997; Stewart, T. A.: 1994) (45).

9. BENEFITS OF GREEN HUMAN CAPITAL

Johnson, W. H. A.: 1999(23) and Dzinkowski, R.: 2000 hypothesized that the human resources of organizations positively affect their upper hands. Besides, companies participating in ecological administration and green advancement effectively can limit the creation of waste and increment efficiency as well as charge generally excessive costs for green items, improve corporate pictures, and in this manner have a beneficial outcome on corporate upper hands under the patterns of famous environmentalism awareness of shoppers and severe worldwide guidelines of natural insurance (Rondinelli, et al. 1998 (4); Chen, Y. S, S.-B. Lai, and C.-T. Wen.: 2006 (7), Porter, M. E., and C. van der Linde: 1995 (35) Shrivastava, P.: 1995) (44).

10. POSITIVE EFFECTS OF STRUCTURAL CAPITAL

Past investigations called attention to that organizations' structural capital positively affects their upper hands (Stewart, T. A.: 1994 (45), Grantham, C. E., and L. D. Nichols: 1997) (14). Furthermore, Porter, M. E., and C. van der Linde: 1995, (35) Shrivastava, P.: 1995, (44) and Rondinelli, et al. (1998) (4) contended be organizations give a lot of consideration and numerous interests in ecological administration and green advancement would not just stay away from the difficulty of fights or discipline about environmental insurance, yet additionally, improve their corporate pictures, upgrade their creation effectiveness, grow new natural business sectors, and consequently increment their upper hands.

11. GREEN RELATION CAPITAL

The Human Resource work had been famous as the driver of association's ecological traditions by putting where else orchestrating its strategies as well as arrangements combined with the objective of manageability Mandip, G. (2012) (27); Cherian, J., & Jacob, J. (2012) (8). Marburg, S. et al (1993). Have convoluted that to get into eco-friendly control, representatives should be revealed from a few out of doors innovative motivation, deliver the location or capability to perform substance. Previous reviews have accentuated the association between manpower administration rehearses as well as superintending corollaries like productivity, versatility, as well as monetary enactment (for example Prennushi, G. et al., (1997) (18); Pillai, R. R. et al., (1999) (31); Clark, K. D. et al., (2003) (10); Yet Laursen also, Foss (2003) have uncovered that very little uncommon significance or noticeable quality has been given on relaying these results to upheaval execution as well as natural administration drives (Redman, T. et al., 2008). Shahbazmoradic, M.R. et al. (2015) (14)figure out that is a necessity towards blending ecological the board in ecological human force administration being that it has conversely vital sooner than just enchanting. Efficacious natural administration in an association needs unique pursuits about manpower resources on the board (Rothenberg, 2003) (41). Furthermore, Jabbour and Santos (2010)(20) likewise speak as gripping natural execution collars require human resource rehearses to help the complete accomplishment as well as protection of natural directing structures in the organizations. organizations, as may arrange manpower administration, rehearses besides goals on ecological leads might define as a flawless point and consequences incorporating ecological leads undertaking Jabbour, C. J. C. (2011)(19). The assessment is done by Harvey et al. (2012) (15) finished up as manpower leads suppose an excellent part in the accomplishment of ecological strategies as well as shows an engagement of human force leads to ecological prosecution. Ecological prosecution might be relying on worker inclusion as well as execution in respect of eco-friendly rehearses with all living space as well as result from the eco-friendly consequences, eco-friendly effects hold an outcome of ecological developments: newest natural drives, newest procedures towards efficacious exertion on assets, effects for squandering lesson, defilements reduces, and such like Goldman, L., Lutz, R. et al., 1993; (5) Ramus, C. A. & Steger, U. (2000) (36); Ramus, 2001 (37); Ramus, C. A. (2001 and 2002 (38)).

12. DATA COLLECTION AND ANALYSIS

13. Newest data is collected through a deliberate idea ballot which means a set of predefined questions. This survey is an element for progress because the result is to deliver to accumulate information on each tested degree in pageant with actual studies arguments. The structural questions became withinside the shape of the Likert scale wherein respondents appraised exceptional declarations concerning self-reliance in addition to relying on determinants. A scale of 1-five became used; the questionnaires had been controlled by the researcher to all of the appellants. The appellants had been given some days to five their proper responses and then they had been gathered for the study.

A scale of one-5 has become used; the questionnaires had been controlled via the investigator to all of the respondents. The respondents had been given one week to fill out the questionnaires and then they have been amassed for evaluation.

13. REPRESENTATIONS:

RELIABILITY EVALUATION

Reliability evaluation is done to check out the indoor tenacity of each assembly via studying Cronbach's alpha coefficient. According to Amphora (2004), an appropriate scale is above 0.7. The results of the reliability take a look at creating for each gather have been precipitated withinside the underneath table

Constructs	Cronbach's alpha	Cronbach's alpha
Green Human Capital	.816	4
Green Structural Capital	.814	4
Green Relational Capital	.813	5
Green HRM	.900	5

It's capable of being a presence from the reliability records desk above for the reason the Cronbach's Alpha differs some of the variety 0.818 and 0.9. This establishes excessive dependability among the gadgets of the compounds. Perceiving all supplies is taken into consideration as useful and constant for calculating scales, no gadgets had been being removed.

14. CORRELATION ANALYSIS

	GHC	**GSC**	**GRC**
EHM	.710	.613	.716
Person correlation			
Sig. ((2- tailed)	.000	.000	.000
N	432	432	432

**. Correlation is vast on the 0.01 level (2-tailed). The consequences of the evaluation of the tool illustrate that all 3 independent determinants are exceptionally substantially related with withinside the based determinants. The previous desk factor speaks about the connection between ecological human force management and green human capital with the values (r =.710, p < 01). Affiliation among ecological structural resource and ecological human force routines is (r =.613, p < 01), and green relational capital is associated with ecological manpower strategies that incorporate the evaluates (r =.716, p < 01). Represents the power association desk with the aid of using Cohen, J. (1988) (9), the affiliation of all self-i.e., unbiased determinants green human capital, green structural capital, green relational capital are overall performance are robust as each correlation coefficient has been better than 0.5.

To extra take a look at of hypotheses, multiple regression evaluation changed into performed to calculate the connection among the determinants, primarily based totally on the connection that subsists among the determinants. This technique especially examines how the 3 impartial determinants are related to the established elements. From the regression table, it can be located that the R-square evaluation is 0.612, because of this that the impartial determinants (i.e., green human capital, green structural capital, and green relational capital) make clear 60% of the deviations (R-rectangular) in green practices. Evaluation of R suggests the established variable pleasant of foretelling. It denotes the multiple correlation coefficients.

The better the evaluation of R, the greater carefully join the independent determinate in the direction of the dependent determinants. In the preview, it may be stated that the R-states became found to be 76.% and factor outs the F value of 106 with a p-value <01 displaying the general implication of the structure the as in the table

It may be discovered that the determinants of green relational capital suggested the very best beta cost of 0.367 in addition to a sizable cost of 0.000 the better rating on this scale indicates a better result on the company's green practices. The aspects of green human capital conduct become discovered to comprise a beta cost of 0.187, with a p-cost of 0.005. The elements of ecological structural capital and improvement are discovered to be reliably sizable in the direction of the organization's overall green performance. Every based variable consists of a sizable favorable affiliation with the firm's green outcomes.

Regression analysis of ecological manpower regulations and ecological intellectual capital

Independent variables	Co-efficient	t-test	p-value
Green human capital	.187	2.834	.005
Green structural capital	.251	3.617	.000
Green relational capital	.367	5.625	.000
R-Square	0.612		
Adjusted R- Square	0.586		
F- Statistics	106.517		

Dependent determinants of ecological manpower regulations

15. OBSERVATION OF THE FOLLOWING

Hypothesis	A variance in green practices	Observations
H1	Green human capital	Positively significant
H2	Green structural capital	Positively significant
H3	Green relational capital	Positively significant

16. CONCLUSION

The necessity of this analysis is to figure out the association between a company's green practices and its economic benefits that are nothing but organizational intellectual capital. We hypothesized the significant connection between these determinants and our hypothesis has been accepted in this analysis. The study helps to find the significance of using green practices to gain intellectual capital. The green practices are not only useful for saving natural resources it is also used for saving the company's resources by effectively using them and it mainly speaks about reducing wastage. And it is having lots of elements, aspects, and perceptions. On the whole, it is useful for the entire system. The amount and variation of the use of green practices may vary due to the industry, place, understanding of the organization, and acceptance or cooperation of employees. Further purposely, the findings point out that business may develop their performance by taking up Green Human resource management.

Reference

1. Alshehhi, A., Nobanee, H., & Khare, N. 2018. The Impact of Sustainability Practices on Corporate Financial Performance: Literature Trends and Future Research Potential. Sustainability, 10(2): 494-519.
2. Arulrajah, A.A., Opatha, H.H.D.N.P., & Nawaratne, N.N.J. 2015. Green Human Resource Management Practices: A Review. Sri LankanJournal of Human Resource Management, 5(1): 1-16.
3. Bangwal, D., & Tiwari, P. 2015. Green HRM – A Way to Greening the Environment. IOSR Journal of Business and Management, 17(12): 45-53.
4. Berry, M. A. and D. A. Rondinelli: 1998, Proactive Corporate Environmental Management: A New Industrial', Academy of Management Executive, 12(2), 38 50.
5. Callenbach, E., Capra, F., Goldman, L., Lutz, R. & Marburg, S. 1993. Eco-Management: The Elmwood Guide to Ecological Auditing and Sustainable Business, Berrett-Koehler, San Francisco, CA.
6. Chari, A., Chen, W., & Dominguez, K.M. 201). Foreign Ownership and Firm Performance: Emerging Market Acquisitions in the United States. IMF Economic Review, 60(1): 1-42.
7. Chen, Y. S, S.-B. Lai and C.-T. Wen.: 2006, The Influence of Green Innovation Performance on Corporate Advantage in Taiwan', Journal of Business Ethics 67(4), 331-339.
8. Cherian, J., & Jacob, J. 2012. A study of Green HR practices and its effective implementation in the organization: A review. International Journal of Business and Management, 7, 25–33.
9. Cohen, J. 1988. Statistical Power Analysis for the Behavioral Sciences (2nd ed.). Hillsdale, NJ: Lawrence Erlbaum Associates, Publishers
10. Collins, J., & Clark, K. D. 2003, Strategic human resource practices, top management team social networks, and firm performance: the role of human resources in creating organizational competitive advantage, Academy of Management Journal, 46, 740–51.
11. Dzinkowski, R.: 2000, The Value of Intellectual Capi talk, The Journal of Business Strategy 2(4), 3? 4.
12. Edvinsson, L. and M. S. Malone: 1997, Intellectual Capital: Realizing Your Co.'s True Value by Finding Its Hidden Roots (Harper Collin Publishers, USA).
13. Emmanuel, J.F. 2015. Financial Sustainability for Non-Profit Organizations. New York: Springer publishing company.
14. Fayyazia, M., Afsharc, Z., & Shahbazmoradic, M.R. 2015. Investigating the barriers of the green human resource management implementation in the oil industry, Management Science Letters, 5, 101–108.
15. Grantham, C. E., and L. D. Nichols: 1997, A Framework for the Management of Intellectual Capital in the Health Care Industry, Journal of Health Care Finance 2(3), 1-19
16. Harvey, G., Williams, K., & Probert, J. 2012, Greening the airline pilot: HRM and the green performance of airlines in the UK. The International Journal of Human Resource Management, 23, 1-15.
17. Hur-Yagba, A.A., Okeji, I.F., & Ayuba, B. 2015, Analyzing Financial Health of Manufacturing Companies in Nigeria Using Multiple Discriminate Analysis, International Journal of Managerial Studies and Research, 3(7): 72-81.
18. Ichniowski, C., Shaw, K. & Prennushi, G. 1997, The effects of human resource management practices on productivity: a study of DOI: 10.9790/487X-171214553 www.iosrjournals.org 52 | Page steel finishing lines. American Economic Review, 87, 291–313
19. Jabbour, C. J. C. 2011, How green are HRM practices, organizational culture, learning, and teamwork? A Brazilian study. Industrial and Commercial Training, 43, 98-105.

20. Jabbour, C. J. C., Santos, F. C. A., & Nagano, M. S. 2010, Contributions of HRM throughout the stages of environmental management: Methodological triangulation applied to companies in Brazil. The International Journal of Human Resource Management, 21, 1049– 1089.
21. Jirawuttinunt, S. 2018, The Relationship Between Green Human Resource Management and Green Intellectual Capital of Certified ISO 14000 Businesses in Thailand. St. Theresa Journal of Humanities and Social Sciences, 4(1): 20-37.
22. Johnson, W. H. A.: 1999, An Integrative Taxonomy of Intellectual Capital: Measuring the Stock and Flow of Intellectual Capital Components in the Firm, International Journal of Technology Management 18(5-8), 562-575.
23. Karaca, S., & Ekşi, İ. 2012, The Relationship between Ownership Structure and Firm Performance: An Empirical Analysis over Istanbul Stock Exchange (ISE) Listed Companies. International Business Research, 5(1):172-181.
24. Lebacq, T., Baret, P.V., & Stilmant, D.2013, Sustainability Indicators for Livestock Farming: A Review. Agronomy for Sustainable Development, 33: 311-327.
25. León, P. 2001, Four Pillars of Financial Sustainability: Resources for Success Series (2). Virginia: The Nature Conservancy.
26. Mampra, M. 2013, Green HRM: Does it help to build a competitive service sector? A study. In Proceedings of the Tenth AIMS International Conference on Management, Bangalore, India, 6–9 January 2013: 1273–1281.
27. Mandip, G. 2012. Green HRM: People management commitment to environmental sustainability, Research Journal of Recent Sciences, 1, 244–252.
28. Masoulas, V.: 1998, Organizational Requirements Definition for Intellectual Capital Management, International Journal of Technology Management 16(1-3), 126? 143.
29. Masri, H. A., & Jaaron, A.A.M. 2017, Assessing Green Human Resources Management practices in Palestinian manufacturing context: An empirical study, Journal of Cleaner Production, 143: 474–489.
30. Mathapati, C.M. 2013, Green HRM: a strategic facet, Tactful Management Research Journal, 2(2): 1-6
31. Mendelson, H. & Pillai, R. R. 1999, Information age organizations, dynamics, and performance, Journal of Economic Behavior and Organization, 38 (1), 253–81.
32. Okoye, L.U., Erin, O.A., Ado, A., & Isibor, A.A. 2017, Corporate Governance and Financial Sustainability of Microfinance Institutions in Nigeria, In 29th IBIMA conference Sustainable Economic Growth, Education Excellence, and Innovation Management through Vision 2020, 3-4 May 2017, Vienna, Austria.
33. Ooi, S. K., Amran, A., Goh, S., & Nejati, M. 2017, Perceived Importance and Readiness of Green HRM in Malaysian Financial Services Industry, Global Business and Management Research: An International Journal, 9(4):457- 474.
34. Opatha, H.H., & Arulrajah, A.A. 2014, Green Human Resource Management: Simplified general reflections, International Business Research, 7(8): 101–112
35. Porter, M. E. and C. van der Linde: 1995, Green and Competitive, Harvard Business Review 73(5), 120-134.
36. Ramus, C. A. & Steger, U. 2000, The roles of supervisory support behaviors and environmental policy in employee "eco-initiatives" at leading-edge European companies, Academy of Management Journal, 41, 605-26.
37. Ramus, C. A. 2001, Organizational support for employees: Encouraging creative ideas for environmental sustainability, California Management Review, 43, 85–105.
38. Ramus, C. A. 2002, Encouraging innovative environmental actions: What companies and managers must do, Journal of World Business, 37, 151–164.
39. Renwick, D., Redman, T. & Maquire, S. 2008, Green HRM: A Review, Process Model, and Research Agenda, Discussion Paper Series, University of Sheffield Management School, The University of Sheffield. http://www.sheffield.ac.uk/content/1/c6/08/70/89/2008-01.pdf.
40. Renwick, D.W., Redman, T., & Maguire, S. 2013, Green Human Resource Management: A Review and Research Agenda. International Journal of Management Reviews, 15(1): 1-14.

41. Rothenberg, S. 2003, Knowledge content and worker participation in environmental management at NUMMI. Journal of Management Studies, 40, 1783-1802.
42. Sa-Dhan Microfinance Resource Centre 2005, Sustainability of Microfinance Interventions, Perspective Paper: 1-20.
43. Sharma, K. 2016, Conceptualization of Green HRM and Green HRM Practices: Commitment to Environment Sustainability, International Journal of Advanced Scientific Research and Management, 1(8): 74-81.
44. Shrivastava, P.: 1995, Environmental Technologies and Competitive Advantage, Strategic Management Journal 16(Special issue), 183-200.
45. Stewart, T. A.: 1994, Your Company's Most Valuable Asset: Intellectual Capital, Fortune 130(7), 68-74
46. Tang, G., Chen, Y., Jiang, Y., Paillé, P., & Jia, J. 2017, Green Human Resource Management Practices: Scale Development and Validity, Asia Pacific Journal of Human Resources, 56: 31–55.
47. Umobong, A. 2015, Assessing the impact of Liquidity and Profitability on Growth of Profits in Pharmaceutical Firms in Nigeria, European Journal of Accounting, Auditing, and Finance Research, 3(10):97-114.
48. Wallstedt, N., Grossi, G., & Almqvist, R. 2014, Organizational Solutions for Financial Sustainability: A Comparative Case Study from the Swedish Municipalities, Journal of Public Budgeting, Accounting, and Financial Management, 26(1): 181-218.

IOP Conf. Series: Earth and Environmental Science 1057 (2022) 012013 doi:10.1088/1755-1315/1057/1/012013

Nutritious diet, a right or a privilege?

Marina Arteaga[1], Mirian Villavicencio[2], Gabriel Brito[3] and Jenny Ulloa[4]

1 Universidad de Guayaquil, Facultad de Ingeniería Química, Guayaquil, Ecuador marina.arteagap@ug.edu.ec, ORCID ID: 0000-0002-5378-9776
[2] Universidad de Guayaquil, Facultad de Ingeniería Química, Guayaquil, Ecuador mirian.villavicenciop@ug.edu.ec, https://orcid.org/0000-0001-6144-9467
[3] Fundación Cleotilde Guerrero, Guayaquil, Ecuador, gabriel.brito@fundacioncleotildeguerrero.org , https://orcid.org/0000-0002-3879-960X
[4] Fundación FUNDAR Ecuador, Guayaquil Ecuador, jenyulloa@gmail.com, https://orcid.org/0000-0002-5763-3576

Abstract. Food security is a right, but in poor countries such as Ecuador, it becomes a privilege because unemployment, the high cost of the basic family basket and the vital family basket limit access to nutritious food. Pearson's statistic was applied to determine if there is a correlation between the study variables, where it was obtained that there is a direct linear correlation between the percentage of undernourished people, GDP and unemployment in Ecuador of 0.28 and 0.31 respectively, and a direct inverse relationship with the value of the basic family food basket of -0.87; currently Ecuador has a 12.4% rate of undernutrition compared to its average for the period 2001-2009, which is due to the implementation of the School Food Program in 1999.

1. Introduction

Food security exists when all people, at all times, have physical, social and economic access to sufficient, safe and nutritious food that meets their daily energy needs and food preferences for an active and healthy life; this concept was defined at the World Food Summit in 1996 (1). The economic crisis of 2020, derived in large part from the AIDS-19 pandemic, triggered one of the largest increases in world hunger in decades, affecting almost all low- and middle-income countries as described in (2).

Food is one of the factors which, with the greatest certainty and importance, condition not only physical development and well-being, but also health, performance, human productivity, the development of communities and their possibilities for improvement in the future. The way of eating and nutritional needs are closely related to the way of life of individuals in their natural environment (3). The first meal of the day is a very important moment from a metabolic and nutritional point of view, because in addition to recharging the muscles with sugar reserves (glycogen) that are used up during the night's rest and providing the body with the energy necessary to face the day, it is a critical moment for the regulation of hunger for the whole day (4).

International treaties signed in the field of human rights require states to adopt effective measures to realize the right to food, but to do so it is essential to increase the level of income, guarantee access to

productive resources and design agrarian and food systems that make it possible to meet the established objectives.
Regarding the need to increase the level of income, it should be emphasized that the lack of food availability is mainly attributable to poverty, not to a shortage of stocks, so that the best way to combat hunger is to do so by increasing the resources of the poorest people(5).

2. Methodology

Pearson correlation statistical method was used, this linear association is applied when the variables are numerical and are crossed in pairs. As the variables are numerical, it first proceeds to see their relationship through a scatter or X-Y graph. Then the covariance (SW) of the variables is calculated and its standardization is Pearson's correlation coeficient (r or rxy). The interpretation of the coeficient r implies a contrast of hypotheses so it can be considered analysis as well as description. The interpretation of r was accompanied by the scatter plot since the relationship should be considered linear. It indicates or measures the linear association or dispersion of the points with respect to an imaginary line (the linear regression line or the line fitted by ordinary least squares [OLS] (6), the study variables are the number of people undernourished, unemployment and vital family hoop.

Figure 1 shows the interpretation of the correlation coefficient obtained, si - ≤ r ≤, where the sign indicates whether the relationship is direct or inverse linear and its value in absolute terms to the intensity of the relationship.

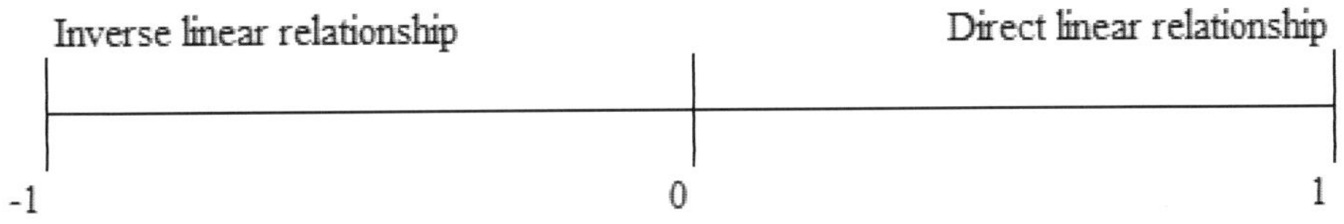

Figure 1. Interpretation of correlation coefficient

3. Results

Health is a state of complete physical, psychological and social well-being, and not only the absence of disease or illness, which should express the right to a full life without deprivation(7). In Ecuador, 25% of undernourished people decreased between 2008 and 2009 but in 2014 it began to grow again and the trend has continued to date.

An inadequate breakfast can lead to the beginning of a vicious cycle that will lead to excessive hunger by mid-morning and, as a consequence, to an easier consumption of high-calorie and unhealthy foods. This, in turn, will lead to inadequate consumption at lunch, due to the abundant snack that has been taken, and so on until dinner, risking compromising the diet for the whole day(4). Ecuador in 1999 implemented the School Feeding Program [PAE] (8), although it is a poor country whose Unified Basic Wage [SBU] is $400.00 in the year 2021, while in the same year the Institute of Statistics and Census [INEC] reported that the cost of the basic family basket stood at USD 715.31 (9) and that of the family basket vita at $502.75; The values of both food baskets are above the value of the unified basic salary, which limits access to healthy food for the population with this type of income, whose malnutrition percentages continue to be low in reference to the 2001-2009 period. Article 13 of the Constitution of the Republic of Ecuador states that the Ecuadorian state, in general, must ensure that people have access to healthy and adequate food, in general terms; also, in addition to guaranteeing the right to food for all people, the state must protect priority groups, due to their quality and vulnerability, directly and without delay and receive specialized attention in the public and private spheres, to protect their constitutional right to receive food (10).

Figure 2 shows that the unemployment and GDP variables there is a direct relationship, but not as intense, since the values are less than 0.50, values were taken from (11) , (12) and (13).

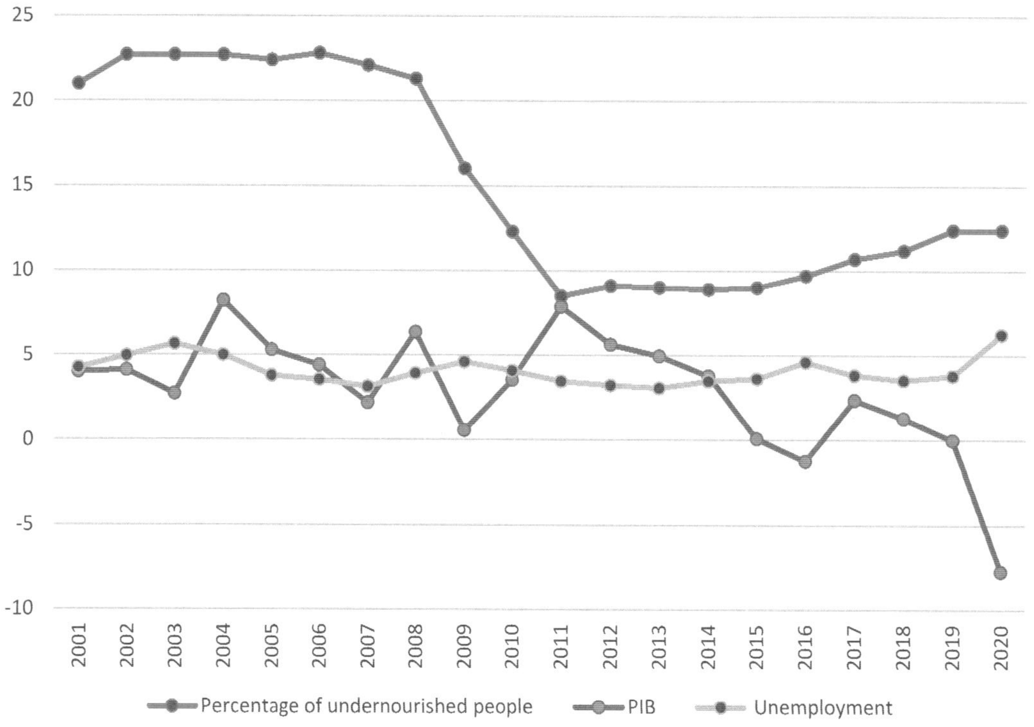

Figure 2. Number of people undernourished & PIB Ecuador & Unemployment

Figure 3 shows the dollar value of the cost vital family hamper in Ecuador.

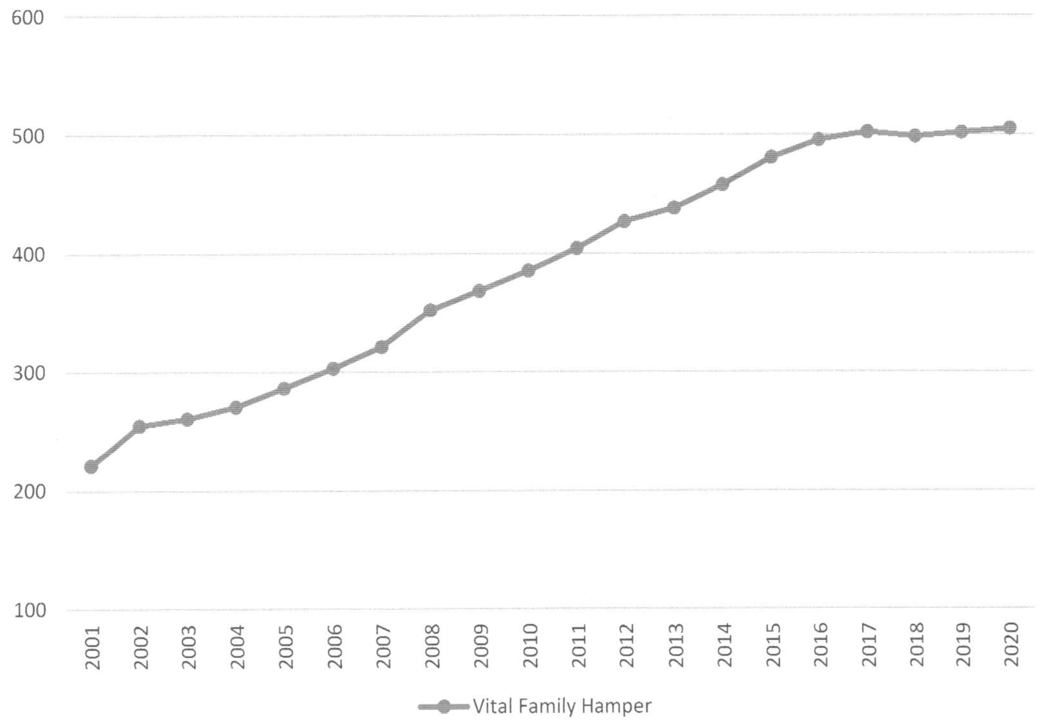

Figure 3. Cost Vital Family Hamper in Ecuador

Table 1 shows the Pearson correlation coefficient obtained for each study variable, percentage of malnutrition 0.28, GDP 0.31 and vital family basket -0.87, this negative value indicates that there is a strong inverse relationship.

Table 1. Pearson Coefficient

	Percentage of undernourished people	PIB	Unemployment	Vital Family Hamper
Percentage of undernourished people	1.00			
PIB	0.28	1.00		
Unemployment	0.31	-0.46	1.00	
Vital Family Hamper	-0.87	-0.53	-0.20	1.00

4. Conclusion

There is a direct linear correlation between the percentage of undernourished people, GDP and unemployment in Ecuador of 0.28 and 0.31 respectively, and a direct inverse relationship with the value of the vital family basket of -0.87. The current percentage of undernourishment of 12.4% is lower than the average for the 2001-2009 period, which is 21.5

References

1. La Seguridad Alimentaria: Información para la toma de decisiones [Internet]. [citado 16 de enero de 2022]. Disponible en: https://www.fao.org/3/al936s/al936s00.pdf

2. El hambre aumenta a nivel mundial durante la pandemia [Internet]. Noticias ONU. 2021 [citado 16 de enero de 2022]. Disponible en: https://news.un.org/es/story/2021/07/1494232

3. López Plana MA. Dietética y nutrición. [Internet]. Málaga: Editorial ICB; 2018 [citado 2 de enero de 2022]. 231 p. Disponible en: https://elibro.net/ereader/elibrodemo/117554

4. OCU Ediciones SA. Cuaderno de alimentación por una vida sana [Internet]. Ediciones de la U; 2019 [citado 3 de enero de 2022]. Disponible en: http://www.ebooks7-24.com/?il=9454

5. Carretero García A. La vulneración del derecho humano a la alimentación: consecuencias de los actuales modelos de producción y consumo de alimentos [Internet]. Madrid: Reus; 2019. 301 p. (Derecho agrario y alimentario). Disponible en: https://bv.unir.net:2769/es/ereader/unir/121445

6. Puente Viedma C de la. Estadística descriptiva e inferencial. [Internet]. Madrid: Ediciones IDT; 2018 [citado 31 de enero de 2022]. Disponible en: https://public.ebookcentral.proquest.com/choice/publicfullrecord.aspx?p=5486569

7. Cuesta U, Bustamante E, Gaspar S, Terrón JL, Peñafiel C. Comunicación y salud [Internet]. Dextra Editorial; 2017 [citado 30 de diciembre de 2021]. 685 p. Disponible en: http://www.ebooks7-24.com/?il=4842

8. kevinrodriguez08. Inicios del Programa de Alimentación Escolar. [Internet]. educa 593. 2016 [citado 31 de enero de 2022]. Disponible en: https://educa593.wordpress.com/2016/01/18/inicios-del-programa-de-alimentacion-escolar/

9. Boletin_tecnico_01-2021-IPC.pdf [Internet]. [citado 31 de enero de 2022]. Disponible en: https://www.ecuadorencifras.gob.ec/documentos/web-inec/Inflacion/2021/Enero-2021/Boletin_tecnico_01-2021-IPC.pdf

10. DERECHO A LA ALIMENTACIÓN - Derecho Ecuador [Internet]. 2018 [citado 31 de enero de 2022]. Disponible en: https://derechoecuador.com/derecho-a-la-alimentacion/

11. 2.1.1 Prevalence of undernourishment | Sustainable Development Goals | Food and Agriculture Organization of the United Nations [Internet]. [citado 31 de enero de 2022]. Disponible en: https://www.fao.org/sustainable-development-goals/indicators/211/en/

12. Crecimiento del PIB (% anual) - Ecuador | Data [Internet]. [citado 16 de enero de 2022]. Disponible en: https://datos.bancomundial.org/indicator/NY.GDP.MKTP.KD.ZG?locations=EC

13. Desempleo, total (% de la población activa total) (estimación modelado OIT) - Ecuador | Data [Internet]. [citado 16 de enero de 2022]. Disponible en: https://datos.bancomundial.org/indicator/SL.UEM.TOTL.ZS?locations=EC

Blue green algal farming revolution: An eco-friendly top-up step for Agribusiness advancements

Parvati Sharma[1], Ankur Kumari[1], Anuradha Bhardwaj[2] Asha Poonia[1] and Yashpal[2] [1]Department of Zoology, Chaudhary Bansi Lal University, Bhiwani (Haryana). [2]National Research Centre on Equine, Hisar (Haryana).

Abstract: Global population growth rate accelerating earth ecosystem disturbance and increasing resource scarcity, malnutrition and poverty. These events challenge socio-economic development of each country and obligate scientific communities to develop such novel approaches to overcome this burden, which not only provide a healthy diet but also engage people in setting up their own business. Therefore, microalgae *Spirulina* is the one of the emerging food source with rich protein content and other nutritional components in adequate amount with diverse health benefits. The small scale investments in microalgae biomass production, familiarizing nutraceutical values and convert in consumable forms is the interesting agribusiness with diversified profits and popularization. *Spirulina* production strategies can be conventional and low-cost effective technologies. These techniques also help in designing new innovative possibilities in modifications and enhancements in the existing agriculture products processing strategies for their better use in different industries. Although most of communities are aware about agribusiness strategies of *Spirulina* but still it is in one the demanding and upcoming bright sector. This article reveals the methods of mass production of *Spirulina* and their significant role in agriculture as well as industrial.

1. Introduction

Expanding population demand is one of the major concerns of every nation especially the food. In this context the raising protein demands is triggering the food companies and agriculture sector to compensate the future demands with best substitute. Therefore the new and more appropriate future protein supplement microalgae *Spirulina* with protein content of 63% dry matter is gaining more and more popularity day by day. *Spirulina* is crammed with all essential nutrients like vitamin-B-complex, vitamin-E, β-carotene, and antioxidants, minerals, chlorophyll, and phyco-cyanobilin [1]. The United Nations world at food conference declared that Spirulina as the best food for the future, and itis gaining popularity now a day's among the wide population [2].

Another important aspect of *Spirulina* biomass production in agriculture is that this is very helpful in the development and maintaining sustainability of ecosystems and environment directly and indirectly. People are not so aware about the theoretical and practical outcomes of the *Spirulina* biomass production as a basic raw material for agribusiness prospective. The initiative of biomass exploitation and the sustained production are common barrier in the use of *Spirulina* agribusiness as a subsystem of the national economy [3].

Therefore, this paper highlighted briefly the nutritional and therapeutic characteristics of *Spirulina* food and its importance for sustainable production systems in agribusiness.

2. Nutritional and therapeutic characteristics of *Spirulina*

Among all plant sources *Spirulina* is rich in high quality proteins, essential amino acids, polyunsaturated fatty acids like linolenic acid (ALA), linoleic acid (LA, 36 percent of total PUFAs), stearidonic acid (SDA), eicosapentaenoic acid (EPA), docosahexaenoic acid (DHA) and arachidonic acid (AA) (AA). Moreover *Spirulina* is a good source of several vitamins i.e.Vitamin-Bcomplex, vitamin C, D and E. Some minerals copper, iron, magnesium, manganese, phosphorus, selenium, sodium and pigments like Chlorophyll a, xanthophyll, beta-carotene, echinenone, betacryptoxanthin, oscillaxanthin, and the phycobiliproteins c-phycocyanin and all ophycocyanin also found in *Spirulina* which provide colour and stability of essential nutritional components as shown in table 1[4].

Table.1 : General composition of *Spirulina*

COMPSOTION	QUANTITY (%)
Proteins	**55-70%**
Carbohydrates	**15-25%**
Lipids	**6-8%**
Minerals	**7-13%**
Humidity	**3-7%**
Dietaryfibers	**8-10%**

Spirulina is a nutraceutical food supplement, but its additional possible health advantages created a lot of interest in recent years as a possible source of medicinal compounds and has been shown inmultiple studies shows its therapeutic values, including antioxidant, immunomo-stimulatory, anti-inflammatory, anticancer, antiviral, and antibacterial properties as shown in fig.1 [5]

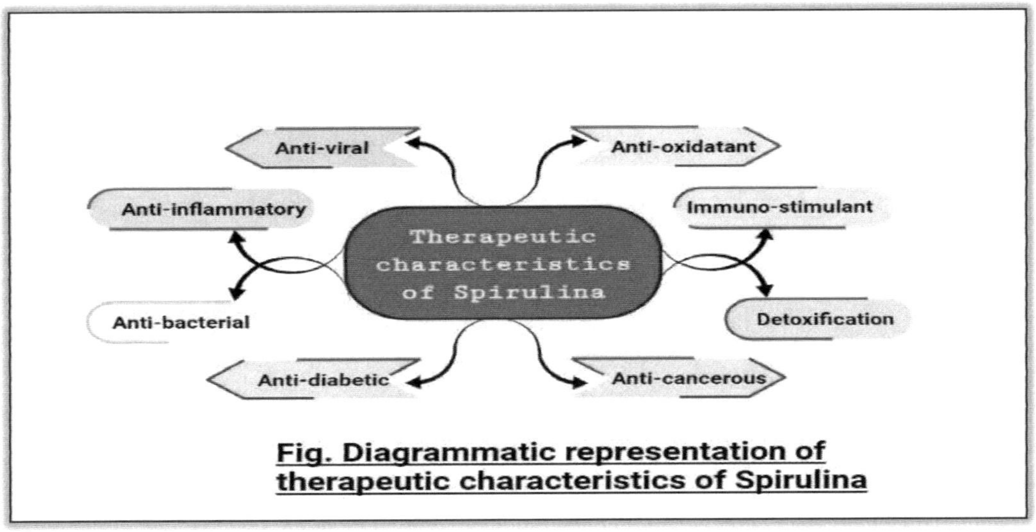

Fig.1Therapeutic characteristics of *Spirulina*

At cellular level a variety of defence mechanisms resist free radical damage including enhanced production of antioxidative enzymes like glutathione S transferase, catalase, and Superoxide dismutases. Spirulina is enriched with dietary components as glutathione, ascorbic acid, polyphenolics, carotenoids, and alpha tocopherol exert antioxidative effects and increases the production of antioxidative enzymes.

3. Commercial production of *Spirulina*

Spirulina is a naturally grown fresh water algae. In natural, habitats *Spirulina* growth cycles can vary as per nutrient supply. Species specific growth can with stand at low temperature15°Cat night and 40°C for a few hours in the daytime. Thus, *Spirulina* cultivation and production is very easy and more convenient & profitable sectors. Many countries are emphasizing the *Spirulina* cultivation as the agribusiness for empowering lifestyle. InJapan, *Spirulina* cultivation is leading than *Chlorella* commercially [6]. Likewise, in 2002 government of India has started a national sector scheme "Agri-Clinics and Agri-BusinessCentre" (ACABC) for empowering youth self-employment by educated them more about agribusiness opportunities. This scheme is also more efficient in giving more extended services to farmer along with their conventional farming methods [7].

The commercial cultivation process of *Spirulina* includes following main steps (as per ACABC process):1) Selection of specific species, 2) Open raceway pond, 3) Maintenance of proper growth conditions, 4) Harvesting system, 5) Drying & grinding and 6) Product processing i.e. capsules or pellets formation as shown in fig.

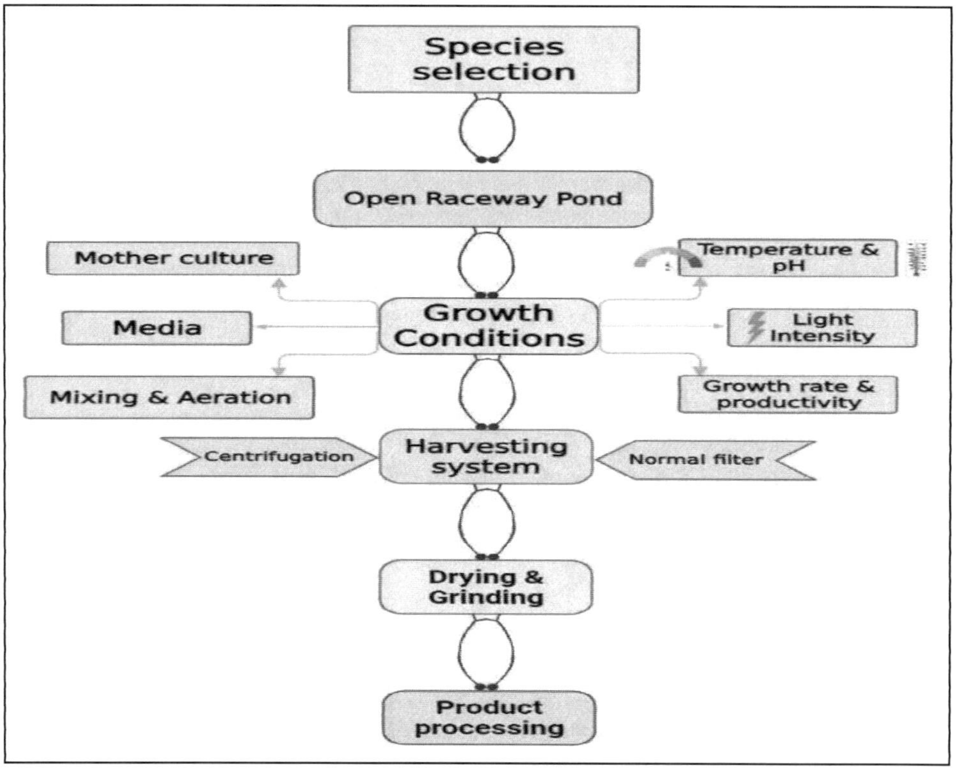

Fig.2: Process of commercial cultivation stages of microalgae *Spirulina*.

3.1 Species selection:

In commercial cultivation process selection of specific species is the most important stage. There are 2 most common species: *Spirulina platensis* and *Spirulina maxima*. Because these species have reported as the positive stimulator, non-toxic effects and contains all essential nutritional components [8].

3.2 Open Race way Pond

The next phase of cultivation is the shallow artificial pond for growing algae. The pond size can vary but depth should be less or and pond can be closed system. The ponds can be concrete ponds or plastic pipes lined large pits [9] & [10].

3.3 Maintenance of proper growth conditions

For proper and increased biomass production of *Spirulina* growth conditions are not so specific. *Spirulina* species uses available nutrients very effectively and increases biomass production manifolds [11]. Some specific conditions like throughout year sunshine, rain, temperature fluctuations,wind etc. affects the growth rate.
The fully developed cultures used for inoculation contains high concentration of coiled filaments which is the indicator of best quality. Growth medium of *Spirulina* culture should contain all essential minerals like sodium carbonate potassium, nitrogen, iron and alkaline water for proper flourishment of culture. For homogenization of culture by proper equal mixing of nutrients is critical step for enhancing productivity [12] & [13]. It can be achieved by rotators, which maintain the cells in suspension by gentle agitation of growing cells. Culture growth with high protein content is between 29°C-35°C [14]. The pH maintenance of the media over 9.0 is obligatory in Spirulina cultures in order to avoid contamination by other algae [15].

3.4 Harvesting system

During harvesting process concentration of algae in pond or the production unit is the major determining factor. Normally after 5 days of seeding process the *Spirulina* culture is ready for harvesting and this is preferred that it should be done during morning time for more protein concentration. This process can be done in 2steps: centrifugation and filteration.

3.5 Drying & grinding

Lightly dried *Spirulina* is more preferred for fresh consumption of *Spirulina*. It is more easily digestible in fresh forms [16]. After drying it can be preserved for longer time by different methods. Normally *Spirulina* is grinded in fine form of average powder size reaches 200-800 nm by grinding 6-10 hrs continuously.

3.6 Product processing i.e. capsules or pellets formation

For marketing and consumer oriented acceptance the powder form is pressed together intotablets and capsules. These forms are formulated as the complete balanced diet with nutrients and immunity boosters [17]. There are some advantages of pellet forms of *Spirulina* as longer shelf life, easily consumable, contains more preservatives and antioxidants and good water stability.

4. *Spirulina* Agribusiness economics:

The *Spirulina* agribusiness is cost effective with less investment as shown on table 2.The pond size should be of 10'x20' and there should be at least 20 ponds for greater production. Approximately 2kg wet culture can be obtained per day and 100 to 130 kg per month from one pond. Thus farmers can earn approximately 40,000-45,000 rupee per month (https://www.agrifarming.in/spirulina-farming-project-report-extraction-process)

Table 2: List of particular items or equipments required for *Spirulina* culture as agribusiness setup (https://www.agrifarming.in/spirulina-farming-project-report-extraction-process)

S.No.	Equipments/Acessories	Price(INR)
1.	Water treatment plant	5000/-
2.	Piping work	25000/-
3.	Electric work	15000/-
4.	Drying screen	10000/-
5.	Miscellaneous	2500/-
6.	Labour (monthly basis)	18000/-
7.	Screens for harvesting	5000/-
8.	Required chemicals (per month)	2000/-
9.	Plant Machinery	15,000/-
10.	Pond construction(20@50,000/-)	10,00,000/-
11.	Materials for packaging	2500/-
12.	Equipments of laboratory	5000/-
TOTALINVESTMENT		**12,50,000/-**
Cost(monthly basis)		**25,000/-**
Income	Spirulina powder sale@600rupeeperkg	**72,000/-**

5. Cultivation training centres in India

Developing countries, like in India, empowering farmers and youths to get more income with less investment. So, government of India initiated training programmes under different schemes as

Microbiology division of ICAR-IARI, NewDelhi- 110012.

GMs Spirulina, C/SNo.121/1,Opposite to Central Admin. Building, Indira Colony,Urunlslampur, Maharashtra – 415 409

Spirulina Production, Research and Training Centre, _ _Kondayampatti village, Madurai Centre for Conservation and Utilization of Blue Green Algae

Nallayan Research Centre for Sustainable Development Navallor village, Kanchipuram district, Tamil Nadu

6. Conclusion

Spirulina is crammed with high protein content than other plant based products and alsocontains all the essential nutrients so it is considered as 'super food'. Many organizations are recommended its salubrious effects on humanwell-being and investing as growing business as wealth and health promising sector. This can be one of the best alternatives of balanced diet to overcome social malnutrition challenges. In developing countries, like India *Spirulina* commercial cultivation at large scale as agribusiness to supply its products to different companies is one of the emerging setup for empowering economic status. Similar, agribusiness has been initiated under ACABC scheme which has great success. It has good turnover with low capital investment, and also providing employment opportunities to others.

7. Future prospective

Scientifically, it has been proven that Spirulina contain the property of antimutagenic, anticancer, antimicrobial, antioxidative, antihyperlipemic, antithrombotic and antihypertensive. Therfore, it works in maintaining good health and longevity and in future it may be include in the food items. Spirulina cultivation strategies should be developed as nature's gift of super food to mankind. According to the report, the global spirulina market is expected to grow at a CAGR of 9.4% from 2019 to reach $629.6 million by 2025. In terms of volume, the spirulina market is expected to grow at a CAGR of 13.6% from 2019 to reach 68,025.2 tons by 2025.

8. References

[1] FathimaKandSalmaP2001.Effect of Spirulina as a NutritionalSupplementonmalnourishedchildren. *Indian J Nutr Diet.*38: 269-273.

[2] Pulz M O and Gross W 2004.Valuable products from biotechnology of microalgae. *Appl. Microbiol. Biotechnol. APP.* 6: 635-648

[3] Hirata T, Tanaka M, Ooike M, Tsunomura T and SakaguchiM 2000. Antioxidant activities ofphycocyanobilin prepared from Spirulina platensis. *Journal of App Phyco.* 12(3-5): 435–439.

[4] Jung F, Krüger-Genge A, Waldeck P, Küpper JH. 2019. Spirulina platensis, a super food? *J. Cell. Biotechnol.* 5(1):43-54.

[5] Hoseini SM, Khosravi-Darani K and Mozafari MR. 2013. Nutritional and medical applications of spirulina microalgae. *Mini Rev Med Chem* 13(8): 1231–1237.

[6] RaviM,LataD S,AzharuddinSandPaulSFD 2010.Thebeneficialeffectsofspirulinafocusingonitsimmune-modulatoryandantioxidantproperties.*Nutr. Diet. Suppl* 2: 73-83

[7] Afroz S, Singh R, Nain M S, Mishra J R, Kumar P, Iquebal M A and Khan S A ProblemTreeAnalysisforDelayinStartingAgribusinessbyTrainedCandidatesunderACABCScheme. *J. Ext. Educ.* 56(2): 22-27.

[8] De Smet PAGM 1997. *CitrullusColocynthis*. In Adverse effects of herbal drugs. *Springer, Berlin, Heidelberg* : 29-36.

[9] Karthikeyan D, Muthukumaran M and Balakumar BS 2016.MassCultivationofMicroalgae in Open Raceway Pond for Biomass and Bio-chemicals Production. *Int. j. adv. res.* 3(2): 247-260.

[10] Singh R N and Sharma S 2012. Development of suitable photo-bioreactor for algaeproduction-areview. *Renew. Sustain. Energy Rev* .16 : 2347-2353.

[11] Sudhakar K, Premalatha M and Rajesh M 2014. Large-scale open pond algae biomassyield analysis in India: A case study. *Int Journal of Sustain Energy* 33(2) :304-315.

[12] Pragya N, Pandey K K and Sahoo P K 2013. A review on harvesting, oil extraction and biofuels production technologies from microalgae. *Renew. Sustain. Energy Rev.* 24: 159-171.

[13] Chisti Y, 2016. Large-scale production of algal biomass: raceway ponds. *In Algae biotechnologySpringer, Cham.*: 21-40.

[14] Kumar M, Kulshreshtha J and Singh G P 2011. Growth and biopigment accumulation ofcyanobacterium Spirulina platensis at different light intensities and temperature. *Braz. J. Microbiol.* 42(3) :1128-1135.

[15] Selvendran D 2015. Large scale algal biomass (Spirulina) production in India. In AlgalBiorefinery:*AnInte App. Springer, Cham.* 151-167

[16] Richmond A and Vonshak A 1978. Spirulina culture in Israel. *Arch. Hydrobiol. Bein.Ergebn.Limnology*. 11: 274-280.

[17] Ogbonda K H, Aminigo R E and Abu G O 2007. Influence of temperature and pH onbiomassproductionandproteinbiosynthesisinaputativeSpirulinasp. *Bioresour. Technol.* 98:2207-2211

[18] Slade R and Bauen A 2013. Micro-algae cultivation for Biofuels: Cost, energy balance,environmentalimpacts and futureprospects. *Biomass & Bioenergy*.53:29-38

IOP Conf. Series: Earth and Environmental Science 1057 (2022) 012015 doi:10.1088/1755-1315/1057/1/012015

Waste management in municipal markets

Diana Merchán [1,2], María Teresa García-Cubero[2,3], Alberto Paredes[4] and María González[5]

1 Universidad de Guayaquil, Facultad de Ingeniería Química, Guayaquil, Ecuador
diana.merchang@ug.edu.ec, ORCID ID: 0000-0002-2352-3838
2 Universidad de Valladolid, Escuela de Ingenierías Industriales, España,
3 Institute of Sustainable Processes, University of Valladolid, Spain
mtgarcia@uva.es, ORCID: 0000-0003-4763-154X
5 Universidad de Guayaquil, Facultad de Ingeniería Química, Guayaquil, Ecuador,
alberto.paredeso@ug.edu.ec, ORCID ID: 0000-0003-1518-4765
6 Universidad de Guayaquil, Facultad de Ingeniería Química, Guayaquil, Ecuador
maria.gonzalezmi@ug.edu.ec, ORCID ID: 0000-0003-1811-6266

Abstract. Municipal markets are approximately 90 years old in Ecuador, although some of them have modern buildings, the process of organic solid waste has not changed. Object of study of this work is the municipal market "Las Manuelas" located in the province of Guayas in the canton of Durán. Organic solid wastes could be used for the elaboration of new biopolymers, so the objective of this research is to determine the environmental impact of the activities articulated to this process. The observation technique was applied to make the process flow diagram and the methodology of the matrix of importance of environmental impacts. Results obtained showed that the activity of preparation of vegetables and fruits within the municipal market presented the highest I with a value of -42, under these parameters the need to establish preventive actions that can contribute to the circular economy was manifested.

1. Introduction

Management of solid waste has become a priority nowadays due to the environmental problems it generates, which has promoted the search for new contingency plans that are at the same time fruitful for the population, as is the case of the reuse of organic solid waste(1). Basic component of the plant wall is lignocellulose, formed mainly by three polymers: i) cellulose, the most abundant, is a homopolysaccharide of β-1,4-glucose; ii) hemicellulose, which follows cellulose in abundance, formed by different branched heteropolysaccharides whose names depend on the type of sugars that form it (xylan, mannan, galactan, glucan, glucuronoxylan, glucuronoxylan, galactoglucomannan, etc.); and iii) lignin, which is a heterogeneous and recalcitrant aromatic polymer, formed by phenylpropane units, that protects the plant against external aggressions (2). Generation and accumulation of waste is not only a serious economic problem, but also has a major environmental impact (3). Municipal markets in the processes of preparation of vegetables and fruits to be placed on hangers, sale of striped cassava, sale of striped corn and sale of plantain per unit, generate organic solid waste that can be used for the development of a biopolymer with properties that contribute to the sustainable development objectives established in 2015 by the United Nations [UNO].

Quality of environmental impact assessment [EIA] processes is directly related to the administrative policies and regulations of each country. In the last decade, EIA has advanced significantly in some Latin American countries, because of increased the technical capacities of experts and improved institutional and regulatory systems for the implementation of environmental impact assessment (4).

Management of organic waste in municipal markets would help reduce solid waste contamination of the soil, water, and air; it also allows for the possibility of economic income from the commercialization of this waste.

Process management implies that a company's or organization's business is conducted under a process approach for the fulfillment of a single objective, and not under departmental objectives. Process is the set of interrelated activities that fulfill an objective. Main part of the procedure is a flujogram or flujo diagram, complemented by information fields such as inputs and outputs, records generated, explanatory notes, codification, version control. (5). Flowcharts are important to establish the beginning and end of a process, its control, graphic analysis, and follow-up.

Proposing and developing measures to implement new policies that facilitate the transition to a circular economy is undoubtedly the first step to achieve it, however, as is evident, all these policies require financing and investment in order to be implemented and to achieve increasingly innovative solutions that respond better to the new problems of our society (6) .The 2030 Agenda for Sustainable Development, which includes 17 Goals and 169 targets, presents an ambitious vision of sustainable development and integrates its economic, social and environmental dimensions. This new Agenda is the expression of the wishes, aspirations, and priorities of the international community for the next 15 years. The 2030 Agenda is a transformative agenda that places the equality and dignity of people at the center and calls for a change in our development style, while respecting the environment (7).

2. Methodology

Participants in the research are the commercial premises of the "Las Manuelas municipal market in Durán canton. For the selection of the sample, the systematic sampling technique was applied and by quotas to ensure that the study was carried out on premises that generate solid organic waste derived from the types of businesses shown in Table 1, which resulted in a quota of 116 premises.

Table 1. Number of commercial premises

No	**Type of business**	**number of stores**
1	Selling Fruits and Vegetables	92
2	Selling striped cassava	4
3	Selling striped corn	4
4	Selling banana per unit	16
	Total	116

Table 2 was used to assess the importance of the environmental impacts identified.

Table 2. Matrix for the evaluation of the importance of environmental impacts

Nature		**Intensity (I)**		**Extension (EX)(Area of influence)**		**Recoverability (MC)(Human Reconstruction)**		**Moment (MO)**	
Beneficial impact (+)	B	Casualty	1	Punctual	1	Immediately recoverable	1	Long-term	1
Beneficial impact (-)	A	Media	2	Partial	2	Recoverable in the medium term	2	Medium term	2
Accumulation (AC)(Progressive increment)		Loud	4	Extensive	4	Mitigable	4	Immediate	4
Simple	1	Very High	8	Total	8	Irretrievable	8	Critical	(+4)
Accumulated	4	Total	12	Criticism	(+4)				
Effect (EF)(Cause-effect relationship)		**Precedity (PR)(Regularity of manifestation)**		**Persistence (PE)(permanence of effect)**		**Reversibility (VR)**		**Synergy (SI)(Regularity of the event)**	
Indirect (secondary)	1	Irregular or aperiodic and discontinuous	1	Fleeting	1	Short Term	1	No synergy (simple)	1
Direct	4	Newspaper	2	Temporal	2	Medium Term	2	Synergistic	2
		Continuous	4	Permanent	4	Irreversible	4	Very synergistic	4

Table 3 shows the significance levels for the negative impacts according to the range, with values between 71 and 92 being of critical importance.

Table 3. Importance assessment for negative impacts

Importance	Absolute value of importance	Color
Irrelevant	< 20	
Low	21 -45	
Moderate	46-70	
Critical	71-92	

Table 4 shows the levels of importance for positive impacts according to the range, with values between 71 and 92 being of critical importance.

Table 4. Importance assessment for positive impacts

Importance	Absolute value of importance	Color
Positive	<45	
Important Positive	46 -92	

Process flow diagram was applied for the graphic analysis of the type of activities that generate solid organic waste in this study.

IOP Conf. Series: Earth and Environmental Science 1057 (2022) 012015 doi:10.1088/1755-1315/1057/1/012015

3. Results

Figure 1 shows the processes that generate solid organic waste, such as coconut husks, corn husks, corn husks and banana stems. It was also identified that the preparation of fruits and vegetables to be placed on the perch generates waste because they remove the first layer of skin so that the product looks fresh or in other cases, they remove the stem and leaves that are not part of the edible part of the fruit or vegetable, for example in the carrot and radish as can be seen in Figure 2.

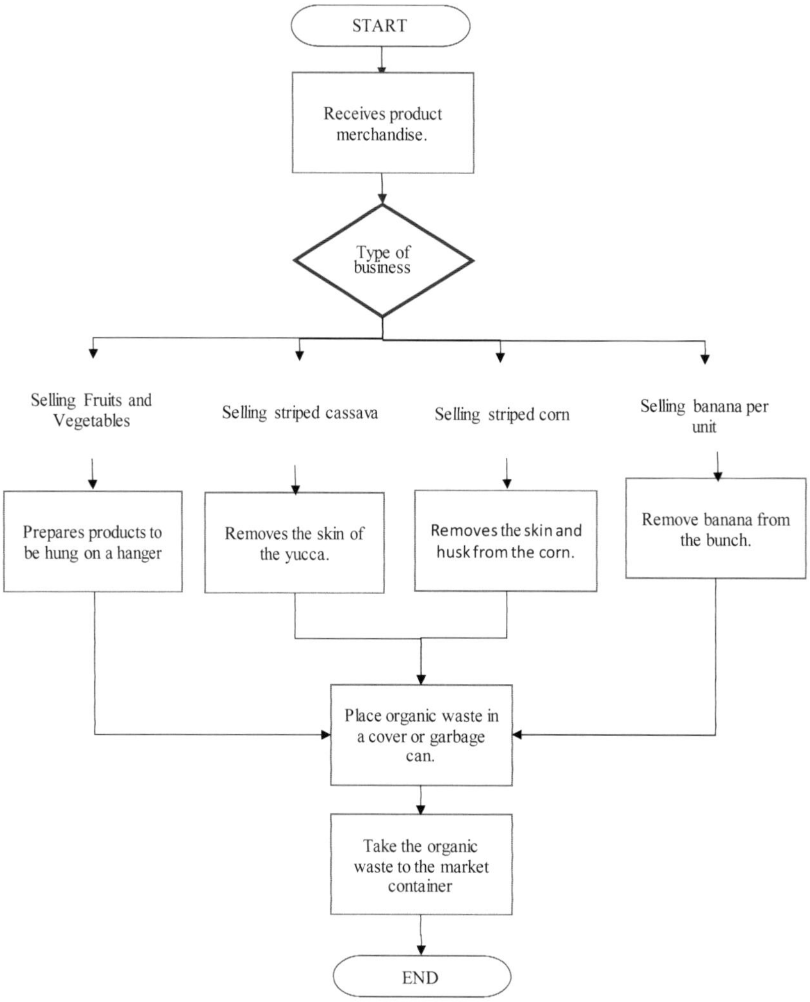

Figure 1 Solid organic waste generation

Figure 2 Carrot and radish.

Process flow diagram of solid organic waste generation shows that solid organic waste has the same destination as other types of waste generated by other commercial establishments in the municipal market. This is an opportunity for the municipal market to apply the principle of segregation (selection or classification) of solid waste at the source of generation, in order to physically separate the waste that is recyclable from that which is not, which facilitates the reuse of materials that can be recycled (8). Considering the type of waste during collection could create additional opportunities, such as facilitating the reuse of waste (9); and thus maintain its constant circulation.

Figure 3 shows the results of the importance of the impacts of the activities that generate solid organic waste, which only generate negative impacts; only the activity of preparing products to be placed on a hanger is of critical importance and the others are of low importance.

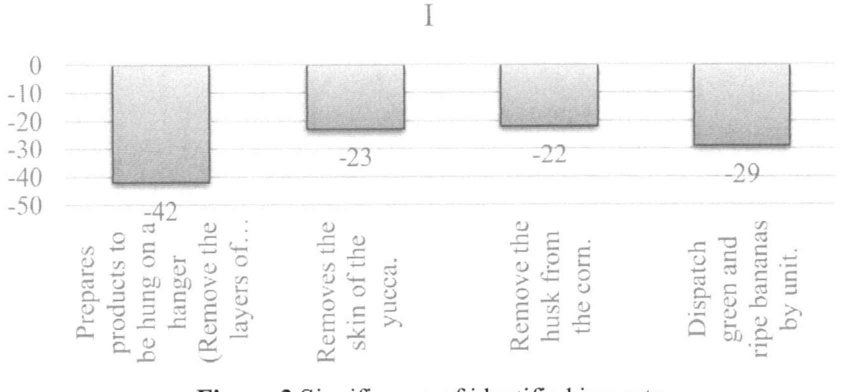

Figure 3 Significance of identified impacts

Through the constant circulation of products and raw materials in the market, wealth can be generated and some of the most important challenges that threaten the entire international community, such as climate change, overexploitation of resources and destruction of ecosystems, can be addressed (6). Giving value to the solid organic waste generated in municipal markets by using it in the development of new products such as the creation of a biopolymer would contribute positively to the actions for the achievement of some of the development objectives defined by the United Nations [UN], among which is to guarantee sustainable consumption and production modalities (10).

4. Conclusion

Activity of preparing vegetables and fruits to be placed on hangers in the municipal market had the highest environmental impact with a value of -42, which places it in the moderate range of importance. In the process flow diagram, it was identified that solid organic waste has the same destination as other types of waste; this situation is a point of opportunity to establish actions to contribute to the circular economy through the commercialization of solid organic waste as raw material for new products such as biopolymer.

References

1. García Alzate LS, Sánchez Tarquino LT, Padilla Sanabria L, Giraldo Giraldo GA. Cellulose Degradation In Household Organic Waste In Social Strata 1 And 2 Using Aspergillus Niger: CELLULOSE DEGRADATION IN ORGANIC HOUSEHOLD WASTE IN STRATA 1-2 BY USING ASPERGILLUS NIGER. Ing Recur Nat Ambiente [Internet]. enero de 2012 [cited 13

february 2022];(11):27-31. Recovered from: https://search.ebscohost.com/login.aspx?direct=true&db=fua&AN=90387404&lang=es&site=ehost-live

2. Martínez MJ, Prieto A. Biotechnology for the utilisation of cellulose and hemicellulose from plant residues. 2021;3.

3. Gañán P, Zuluaga R, Castro C, Restrepo-Osorio A, Cock JV, Osorio M, et al. Cellulose: an evergreen polymer with a bright future. Rev Colomb Mater [Internet]. 28 de agost de 2017 [cited 14 february 2022];(11):1-4. Recovered from https://revistas.udea.edu.co/index.php/materiales/article/view/328779

4. Soto NHS, Barrera VCS, Pérez SCA. Comparative analysis of environmental impact assessment methods applied in the road sub-sector in Colombia.. Rev Investig Agrar Ambient [Internet]. 5 july 2018 [cited 12 de february de 2022];9(2):281-94. Recovered from https://hemeroteca.unad.edu.co/index.php/riaa/article/view/2174

5. Pardo Álvarez, JM. Process and operational risk management. [Internet]. Madrid: AENOR - Asociación Española de Normalización y Certificación; 2017 [citado 14 febrero 2022]. Recovered from http://public.ebookcentral.proquest.com/choice/publicfullrecord.aspx?p=5190227

6. Belda Hériz I, ProQuest. Circular economy: a new model of sustainable production and consumption. Madrid: Editorial Tóbar Flores; 2018.

7. Bárcena A Naciones Unidas, Economic Commission for Latin America and the Caribbean, Spanish Agency for International Development Cooperation P of CTC. The 2030 Agenda and the Sustainable Development Goals: an opportunity for Latin America and the Caribbean. [Internet]. Santiago, Chile: Economic Commission for Latin America and the Caribbean (ECLAC); 2018 [cited 25 Feb. 2022].. Recovered from https://bibliotecadigital.aecid.es/bibliodig/es/consulta/registro.do?control=ES-MAAEC20190011211

8. García Batista RM, Socorro Castro AR, Vanessa Maldonado A, García Batista RM, Socorro Castro AR, Vanessa Maldonado A. Environmental management and management of solid waste, case study. Rev Univ Soc [Internet]. March 2019 [cited 2022 Feb 25];11(1):265-71Recovered from http://scielo.sld.cu/scielo.php?script=sci_abstract&pid=S2218-36202019000100265&lng=es&nrm=iso&tlng=en

9. Mejía C, Llivisaca C, Astudillo F, Vázquez-Rodas A. Automatic collection and sorting of recyclable waste. Maskana [Internet]. 30 December 2017 [cited 25 February 2022];8:331-40. Recovered from https://publicaciones.ucuenca.edu.ec/ojs/index.php/maskana/article/view/1992

10. Development Goals [Internet]. UN. 2015 [cited 25 February 2022]. Recovered from https://guatemala.un.org/objetivos-de-desarrollo/

Biodegradation of Plastic Using Termites and their Gut Microbiota: A Mini Review

Amit Kumar[1], C. M. Kalleshwaraswamy[2], Radhika Sharma[3], Parvati Sharma[1], Asha Poonia*[1]

[1]*Department of Zoology*, Faculty of Life Sciences, Chaudhary Bansi Lal University, Bhiwani, India.
[2] Department of Entomology, College of Agriculture, University of Agricultural and Horticultural Sciences (UAHS) Shivamogga, Karnataka, India
[3]*Department of Biology and Environmental Sciences, CSK HPKV Palampur, India.*

Email: *asha.poonia@cblu.ac.in*

Abstract: Nature's pollution is human created activity and human crosses the entire barrier to causing pollution and posing a high significant risk to health of all animals and also for plants. Plastics are the one of the major reason behind this. The high use of plastics material create high solid waste and this solid waste having no completely and fast degradation methods are available in nature. Their degradation also releases various harmful toxic substances which are harmful for environment. Various methods are available for treatment of plastic such as photo-oxidative degradation, thermal degradation, ozone-induced degradation, mechano-chemical degradation, catalytic degradation, and bio degradation. All these methods pose threat to the environment except the method of Biodegradation. Degradation of plastics using microbes are very eco-friendly degradation and causing no any side effect on nature. Termites are the major soil insect that are also capable to degrade plastics using their gut microbiota. Lot of microbial diversity present in termites gut but only few having potential to degrade plastics. In this review we mainly focus on the diversity of termites and their gut microbial fauna that having potential to degrade plastics and their different polymers.

Keywords: Abiotic factor, biodegradation; gut microbiota; microbes; plastics; plastic degradation; termites.

1. Introduction

In 21st century plastics are used as essential materials due to their physical and chemical properties. Plastics are the first choice as raw materials for every individual for their use and their business. Approximately about 57 million tons of plastics wastes are generated annually in worldwide [1, 2]. Plastics are synthesized by small molecule polymerization (poly-addition or poly-condensation). They are synthetic or semi-synthetic materials and synthesized from organic materials, such as wood, paper or wool, crude oil and renewable raw materials. Plastics are divided into two groups i.e. thermoplastics and thermoset plastics. Thermoplastics are soft when heating and hard after cooling and they are reversible and recycled material. Thermoplastics materials can be found in products such as bottles, films, cups, and fibres. Thermoset plastics cannot melt after setting process. They are irreversible and cannot recycle such as electronic chips, dental fillings and the lenses of glasses. Plastics also used in food and medicine packaging, wrapping, stationary paraphernalia, decoration. High consumption of plastics for use create waste and due to the absence of their auto-degradable properties and less degrading nature, plastics can pose a significant risk of environmental pollution and release very toxic chemicals that are harmful to the environment[3,4]. Biotic agents like various (microbes, insects and their associated enzymes) play a key role in environments cleanness and plastic degradation. They clean environmental contaminants by degrading plastics and their harmful contaminates. Enzymes enhance the rate of reaction of degradation process of plastics and convert plastic polymers into oligomers. In plastic biodegradation fungal species acts as best plastic degrading microorganisms or biotic agent.

Termites also have potential to degrade plastic or their polymers [5, 6]. Generally molecular identification is used in addition to morphological identification for termite's identification [7]. Termites are also called the ecosystem engineers as they convert tough plant biomass into soil. Termites primarily attack moisture-bearing plant materials such as cellulose, hemicellulose, and lignocellulose for food, but also attack non-cellulosic materials, such as plastics, in search of food [8, 9]. Various microbial population harbor in the gut of termites which can help to degrade the plastics polymers [6, 9] Mainly wood feeding higher termites attack on plastics and degrade them via their mandibles. Microbial degradation occurs by both aerobic and anaerobic process. Plastic degradation also depends upon their polymers from which they synthesized. Microbes or termites mainly attack Low density polyethylene (LDPE) and Wood plastic composite (WPC) and recycled high-density polyethylene (HDPE), Polyhydroxyalkanoates, Polylactic acid, Polyethylene terephthalate, Polyester, Polyvinyl alcohol, Nylon, Polyethylene succinat, Polylcaprolactone, Starch/polyester blends, Starch/polyethylene blends, Starch/PVA Blends. So far these polymers bio-degradation was reported [10].

1.1. Low Density Polyethylene (LDPE):

Low density polyethylene (LDPE) is one of the main reason behind the environment pollution as it is very frequently used. Low density polyethylene (LDPE) are constituted from ethylene polymerization. Low density polyethylene has different properties such as opacity, tensile strength, tear strength, hardness and chemical resistance, ductility even at low temperatures. They are incompletely crystalline solids (contain 50–60% crystallinity. LDPE are two types mainly: linear low density polyethylene (LLDPE) and branched low density polyethylene (BLDPE). The main differences between these are density, degree of branching and availability of functional groups on the surface. Its versatile features and utility encourage almost all industrial, agricultural or domestic markets for their use. Every Year around 500 billion to 1 trillion plastic bags are consumed by whole world in which only 1% of these plastic bags are recycled every year [11, 12]. Increasing use of these plastics day by day create a great threat on environments. For reducing the threat of their increasing consumption their degradation are very essential.

IOP Conf. Series: Earth and Environmental Science 1057 (2022) 012016 doi:10.1088/1755-1315/1057/1/012016

1.2. High-density polyethylene (HDPE):

Due to its durability, light weight, low cost, easy procedural effectiveness, and other adaptability, high density polyethylene (HDPE) is a type of synthetic plastic widely utilised in the textile industry, automobile parts, food goods, and laboratory equipment business [13]. It has less branching and more linearity than low-density polyethylene (which has random branching and poor chain packing density), allowing it to have a higher packing density [14]. Polyethylene is a xenobiotic that is resistant to natural breakdown. Furthermore, its hydrophobic nature causes it to degrade slowly [15, 16]. Polyethylene garbage is often disposed of in landfills or dumped into bodies of water to decompose/degrade [17]. Because it accumulates at a pace of 25 million tonnes per year, scientists face a significant difficulty in degrading it [15, 18]. Polyethylene recycling is one of the most environmentally friendly options.

Currently, only a small percentage of plastics are recycled, with the remainder ending up in landfills [19]. Landfills are scarce and often appropriate, whereas burning generates very harmful emissions that pollute the air [20]. Hydrogen chloride, dioxin, cadmium, and fine particle matter are all possible hazardous pollutants from incinerating polyethylene. Furthermore, chlorinated plastics can emit hazardous chemicals into the surrounding soil, which can harm creatures if the pollutants reach ground water or other nearby water sources.

Biodegradation, on the other hand, may be the best option for reducing polyethylene contamination in the environment [21, 22]. Biodegradation is a biochemical process in which microorganisms release enzymes that act as a catalyst. So far, the diversity of microorganisms capable of degrading polyethylene has been limited to 17 bacteria species and 9 fungal genera [23]. Carbon dioxide and water are the end products of aerobic metabolism [24]. Whereas carbon dioxide, water, and methane are the end products of anaerobic metabolism [24,25].

Table1. Types of plastic polymers and their application [26, 27, 28].

Types of Plastic	Applications
Polyethylene terephthalate (PET)	Drinking bottles, peanut butter jars, plastic films, micro-wavable packaging
Polylactic acid (PLA)	A biodegradable thermoplastic generated from lactic acid that can be transformed into a range of aliphatic polyesters.
Polyetherimide (PEI)	Does not crystallize at high temperature
Phenolics (PF) or (phenol)	High-modulus, heat-resistant, and fire-resistant polymer used to insulate portions in electrical fittings, paper laminating items (e.g., Formica), and thermally insulating foams.
Polyvinyl chloride (PVC)	Shower curtains, window frames, and flooring are all examples of plumbing pipes and guttering.
Polyethylene (PE)	A wide range of low-cost applications exist, including supermarket bags and plastic bottles.
Polyvinylidene chloride (PVDC)	Food packaging
Low-density polyethylene (LDPE)	Outdoor furniture, sliding, floor tiles, shower curtains, clamshell packaging

Polyamides (PA)	Fibers, tooth brush bristles, fishing line, under- the-hood car engine mouldings
Polypropylene (PP)	Bottle caps, drinking straws, yogurt containers, appliances, car fenders (bumpers) plastic pressure pipe systems
Polycarbonate/ Acrylonitrile butadiene styrene (PC/ABS)	A stronger plastic made from a combination of PC and ABS. Car interior and exterior pieces, as well as mobile phones, use this material.
Polystyrene (PS)	Food containers, plastic dinnerware, disposable cups, plates, cutlery, CD and cassette boxes and packaging foam.
Melamine formaldehyde (MF)	One of the aminoplasts that can be used as a multicoloured substitute for phenolics. Thermoplastic made of modified corn starch that is biodegradable and heat resistant.
Acrylonitrile butadiene styrene (ABS)	Electronic equipment cases (e.g., computer monitors, printers, keyboards) drainage pipe
High impact polystyrene(HIPS)	Refrigerator liners, food packaging, vending cups
Polyester (PES)	Fibers, textiles
Nylon	Bearings, speedometer gears, windshield wipers, water hose nozzles, football helmets, race horse shoes, and cell phones.
Polytetrafluoro ethylene (PTFE)	Electronics bearing, nonstick kitchen utensils.
Polymethylmethacrylate (PMMA)	Vehicle rear light coverings, contact lenses, glazing, aglets, florescent light diffusers
Urea formaldehyde (UF)	Wood adhesive and electrical switch housings
Polyether etherketone (PEEK)	Thermoplastic biocompatibility with strong chemical and heat resistance allows its usage in implant applications and aerospace manufacturing.

2. Termite's Potential in Plastics Polymer Degradation

Plastic polymers are not completely auto-degradable. They create great risk to nature like plastic bags reduces the soil fertility, depletion of underground water source; they create health problems of both human and grazing animals and pollute the soil, water and air [29, 30]. Wood plastic composite (WPC) and recycled high-density polyethylene (HDPE) are the polymer of plastics mainly consist of wooden material with petrochemical plastics material. In WPC composition 80% lignocellulose particles while 20% fossils based thermoplastics are used. Thermoplastics composition contains polypropylene (PP), polyethylene (PE) and polyvinylchloride (PVC) [31]. Recycled high-density polyethylene and wood both are easily used in WPC formation. WPCs are total bio based plastics and easily biodegrade. Numerous living (fungi, bacteria) and non-living agents (sunlight, moisture and temperature) degrades the WPCs. Termites also degrades the WPCs and recycled high-density polyethylene (HDPE). If the WPCs come in contact with high relative humidity or water so their moisture absorption invites fungi and termites for their destruction. Termites need moisture for degrading their food sources. WPCs surface properties are accelerated by termites as well as weather condition. When weather accelerated various cracks are produces over the surface of WPCs and these cracks provide path for termites. Termites by using their mandibles change the physical or chemical properties of WPCs. They remove the plastics from the surface and getting their food source like wooden materials present in the inner side of the WPCs [5]. Higher termites *Nasutitermesnigriceps* species (as biotic agent) were used for test the affect of termites on WPCs. The richer layer of polymers of plastics make more resistant to termites attack as well as accelerated weathering. Termite not directly attacks the polymer matrix sections. They remove the polymer matrix sections and get their food in the form of wood which are buried inside the polymer layers. According to one of the studies it was found that in WPCs degradation approximately 90% termite's population died and only 10% remained alive [5]. Termites mortality is one of the major reason that the chemical additive is added in WPCs as it act as repellent of termites. Formation of the fault zones also attract termites and provide entrance to interior of the composite. *Cryptotermes brevis* and *Nasutitermes corniger* also have potential to degrade wood boards and thermoplastics [32, 33, 34]. *Coptotermes curvignathus* also degrades WPC when no other choice of food is available [35].

Termites also have capability to degrade buried high-density polyethylene (HDPE) pipes which are mainly used for water and gas transport. After long period of their uses their chemical and mechanical properties are changed by biological agents like termites. Due these biological agents abnormal leakage occurred and loss of their strengthens and other properties [36]. Lenz *et al.*, 2011 did a comparative study demonstrating the ability of subterranean termites (*Coptotermes sp. Mastotermes sp.* and *Reticulitermes sp.*) of different world regions to damage a set of four different plastic materials. It was found that termite species of Australian termites (*C. acinaciformis* and *M. darwiniensis*) damage plastic samples far more readily than any other species investigated [38]. In a study Thamil (2016) reported that primitive dry wood termites were capable of chewing and eating polyethylene carrier bags. In this study two bacterial strains that are capable of digesting polythene were isolated from termite gut and identified as *Lysinibacillus sp.* and *Bacillus sp.* [38]. Kato *et al.*, 2000 patented biologically pure culture of *Pseudomonas cepacia* strain obtained from termite's gut which is capable of degrading trichloroethylene [39]. Terzi *et al.*, 2009 conducted a study to determine the resistance of termites to particle board containing waste tyre rubber it was found that Formosan termites (*Coptotermes formosanus*) were able to degrade the particle board specimens [40].

Table 2. Plastic degrading termites species

Sr. No.	**Termites species**	**References**
1.	*Nasutitermes nigriceps*	[32, 33]
2.	*Cryptotermes brevis*	[34]
3.	*Nasutitermes corniger*	[34]
4.	*Coptotermes formosanus*	[40]
5.	*Coptotermes acinaciformis*	[37]
6.	*Mastotermes darwiniensis*	[37]

IOP Conf. Series: Earth and Environmental Science 1057 (2022) 012016 doi:10.1088/1755-1315/1057/1/012016

7.	*Coptotermes curvignathus*	[35]

2.1. Role of Termites Gut Microbiota in Plastic Degradation

Termites mainly depend on their gut microbial strains for degradation of tough material. So far only bacteria and fungi of termites gut having capability to degrade plastics. Fungal strains having high potential to degrade plastic polymers as compare to bacterial strains. In termites gut two bacterial strain were isolated which having potential to degrades LDPE. These are *Bacillus cereus* and *Lysinibacilluset* [38]. These bacterial colonies are stimulated on the surface of polyethylene material such as polyethylene bags. They are having capacity to grown in both aerobic and anaerobic condition. *Bacillus sp.* are gram positive and rod shape facultative anaerobic strains while *Lysinibacillus* are gram positive and rod shaped spore forming bacteria. *Bacillus sp.* Having potential to degrade about 24.7% of polyethylene bags 30 day while *Lysinibacillus* degrade 27.8% polyethylene in 30 day. Both strains were identified by 16S rRNA gene sequences [38]. Bacterial degradation of polyethylene were analysed by Kathiresan method [29].

Fungal species *Xylaria* that show close exo-symbiosis with termites and found in wood feeding higher termites garden/comb [41]. *Xylaria* sp. also helpful in degradation of plastics polymer sheets and uses plastics sheet as a carbon source [9]. *Xylaria* sp. degrades plastic in both aerobic and anaerobic condition and produce CO_2 and H_2O during aerobic environment while produce CO_2, H_2O and methane in anaerobic breakdown process [9, 42, 43].

3. Techniques Used For Plastic Degradation Analysis

Several techniques were used for plastic degradation analysis such as DMR (Dynamic Mechanical Analysis), SEM, HT-GPC (High Temperature Gel Permeation Chromatography), Thermogravimetric analysis, DSC, X-diffraction, Small and Wide angle X ray Scattering, Fourier Transformed Infra-red Spectroscopy (FTIR), Thin layer Chromatography (TLC), Gas Chromatography (GC), Nuclear Magnetic Resonance (NMR), Matrix Assisted Laser Desorption Ionization-Time Of Flight, Chemilluminesence, Gas Chromatography-Mass Spectrometry (GC-MS), Adenosine triphosphate (ATP), Fluorescien Diacetate (FDA) and protein analysis. The biodegradation of plastics can also be monitored by CO2 evolution rate of CO_2, uptake of O_2, change in chemical properties of polymer , change in physical properties of polymer and the growth rate of organisms [44,45]. Several tests should be followed to evaluate plastic degradation for the following reasons [44].

The leaching of additives, including plasticizers, can result in weight loss.

1. The production of carbon dioxide can result in the low molecular weight part of the polymer degrades, whereas long chains do not.

2. Loss of additives or chemical makeup. Very small changes in plastics can affect the strength of plastics. Several techniques are available to investigate the degree and nature of degradation.

Table3: Techniques used for plastic degradation analysis [45].

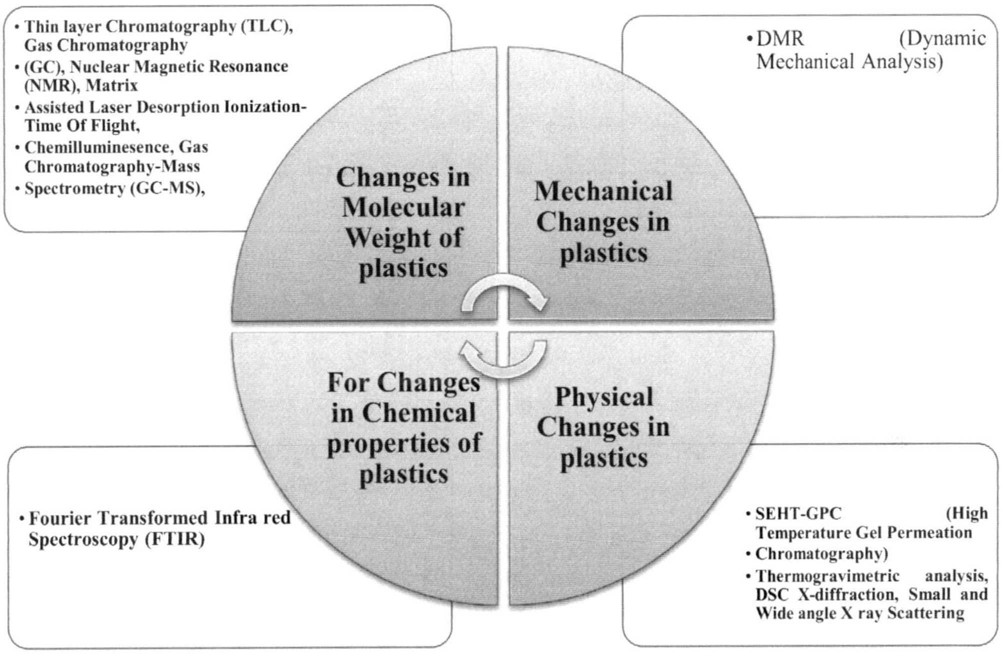

4. Role of Abiotic Factors in Plastic Degradation

Like biotic factor several abiotic factors also change the physical and chemical properties of plastics and degrade them. In abiotic factors mainly sunlight, moisture, temperature, pH and salinity are affects the plastic biodegradation [22]. All environmental abiotic factors not only affect the polymers characteristics but also change the microbial or enzymatic activity [45]. High molecular weight polymers are more resistant to biotic agents as compare to low molecular weight polymers. High molecular weight resists the plastics from biodegradation. While low molecular weight is more easily biodegrade [46, 47]. Melting point are also affects the plastic biodegradation. Temperature of any reaction directly alters the biodegradation process. Temperature also affects the microbial growth. All microbes grow at different specific temperature. All enzymes also activates at specific temperature. If any plastic polymer having low melting points it favours the biodegradation while high melting point resist to plastic from biodegradation because high melting point decrease the enzymatic activity and increase the enzymes degradability [48].

pH also affect the microbial and enzymatic activity. All enzymatic reactions mainly depend on the pH of any compound. Change in pH can easily change the rate of degradation in acidic or basic environment. Various microbial growths are mainly pH dependent [49]. They grow at optimum pH.

So change in pH directly affects the growth of any microorganisms. Plastics shape and size also resist them from biodegradation. Large surface area plastics favour biodegradation as compare to low surface area [50]. High moisture contents in plastics attract biotic agents (like termites) for degradation. Termites need moisture for degradation if polymers having high moisture then termite used them as food source while moisture free plastics resist from biodegradation [51].

5. The impact of termites and their gut bacteria on the environment

We all know that plastics are extremely hazardous to environmental health when they burn because they emit a variety of poisonous chemicals into the environment that impact both living and non-living species. Degradation of plastics through various procedures is always expensive for everyone. Biodegradation processes are therefore particularly successful in minimising plastic garbage or waste. Termites are important in the biodegradation of plastics because they have a large variety of microbial machinery in their guts that may digest plastics and solid waste, lessening environmental impact.

6. Conclusion

As of now we use plastic as an essential product in our daily lives. But excessive consumption of plastic poses a great risk of pollution to our environment. Bio degradation methods are used for plastic degradation as they are considered to be completely eco-friendly. There are many plastic degenerative microbes or other biological agents such as termites, bacteria and fungi present in the environment that play an important role in plastic degradation or clean environments. This study reviewed that termites also have potential to degrade plastics. Termites gut microbiota or their extracellular symbiotic agents play a critical role in plastic or other solid waste degradation. Termite's Gut microbiota have a lot of enzymatic diversity that has the power to degrade plastic and convert it into CO2 and H20. This review concludes that plastic polymers that have very few auto-degradable characters are easily degraded by microbial community of termites. At this hour we need to look for nature friendly technologies that can completely degrade the plastic. Also we also need to pay attention to the minimal use of plastics that cannot recycle or degrade and are harmful to the environment.

Acknowledgement

The authors acknowledge the financial help received from Minor Research Grant provided by Chaudhary Bansi Lal University, Bhiwani, India for this work.

Conflict of Interest

The authors have no conflict of interest.

REFERENCES

1. Bollag W B, Jerzy D, Bollag J M (2000) Biodegradation and encylopedia of microbiology. *In J. Lederberg (ed.). Academic, New York*: 65:461-471.

2. Reddy R M (2008) Impact of soil composting using municipal solid waste on biodegradation of plastics. *Indian J Biotechnol* :7: 235–239.
3. Saminathan P, Sripriya A, Nalini K, Sivakumar T and Thangapandian V (2014) Biodegradation of plastics by Pseudomonas putida isolated from garden soil samples. *J Adv Bot Zool*: *1*(3), 34-38.
4. Das M P and Kumar S (2014) Microbial deterioration of low density polyethylene by *Aspergillus* and *Fusarium* sp. *Int J Chem Tech Res*: *6*(1), 299-305.
5. López-Naranjo E J, Alzate-Gaviria L M, Hernández-Zárate G, Reyes-Trujeque J, Cupul-Manzano C V and Cruz-Estrada R H (2013) Effect of biological degradation by termites on the flexural properties of pinewood residue/recycled high-density polyethylene composites. *Journal of applied polymer science*. *128*(5), 2595-2603 https://doi.org/10.1002/app.38212
6. Yang J, Yang Y, Wu W M, Zhao J and Jiang L (2014) Evidence of polyethylene biodegradation by bacterial strains from the guts of plastic-eating waxworms. *Environmental science and technology*: *48*(23), 13776-13784. https://doi.org/10.1021/es504038a
7. Poonia A and Sharma V (2014) Genetic relationship of two species of odontotermes based on dna sequences of mitochondrial 12s ribosomal rna gene. *International Journal on Environmental Sciences* : Vol. 5 **2** pp 139-147, 2014
8. Kumar A, Dalal J, Poonia A, Kumari A, Lata P, Sharma R and Verma N (2022) Cost effective and sustainable plastics bio-degradation using microbiota. *Shodhsamhita* IX. 80- 95.
9. Kumar A, Poonia A, Sharma R, Jangra M, Sehrawat R and Sansanwal R (2020) Termite gut: home to microbiome. *Uttar Pradesh Journal of Zoology*, 9-23.
10. Ahmed T, Shahid M, Azeem F, Rasul I, Shah A A, Noman M and Muhammad S (2018) Biodegradation of plastics: current scenario and future prospects for environmental safety. *Environmental Science and Pollution Research: 25*(8) 7287-7298 DOI : https://doi.org/10.1007/s11356-018-1234-9
11. Yakowitz H (1990) Incineration of municipal solid waste: scientific and technical evaluation of the state-of-the-art by an expert panel. *Resources, Conservation and Recycling*: *4*(3), 241-251. https://doi.org/10.1016/0921-3449(90)90005-OGet
12. Curlee T R and Das S (1991) Identifying and assessing targets of opportunity for plastics recycling *Resources. conservation and recycling*: *5*(4), 343-363. https://doi.org/10.1016/0921-3449(91)90012-D
13. Begum MA, Varalakshmi B, Umamagheswari K (2015) Biodegradation of Polythene Bag using Bacteria Isolated from Soil. *Int J Curr Microbiol App Sci:* 4(11):674–680
14. Arutchelvi J, Sudhakar M, Arkatkar A, Doble M, Bhaduri S, Uppara VP (2008) Biodegrdation of polyethylene and polypropylene. *Ind J Biotechnol* :7:9–22.

15. Orhan Y, Büyükgüngör H (2000) Enhancement of biodegradability of disposable polyethylene in controlled biological soil. *Int Biodeterior Biodegrad:* 45:49–55. https://doi.org/10.1016/S0964-8305(00)00048-2
16. Hadad D, Geresh S, Sivan A (2005) Biodegradation of polyethylene by the thermophilic bacterium Brevibacillus borstelensis. *J Appl MicrobiolL:* 98(5):1093–1100 https://doi.org/10.1111/j.1365-2672.2005.02553.x
17. Priyanka N, Archana T (2011) Biodegradation of polythene and plastic by the help of microorganisms: a way for brighter future. J Environ Anal Toxicol: 1:111. doi:10.4172/2161-0525.1000111
18. Sangale M K, Shahnawaz M, Ade AB (2012) A Review on biodegradation of polythene: the microbial approach. J Bioremed Biodeg: 3:164. doi:10.4172/2155-6199.1000164
19. Bhardwaj H, Gupta R, Tiwari A (2013) Communities of microbial enzymes associated with biodegradation of plastics. *J Polym Environ*: 21(2), 575-579. doi:10.1007/s10924-012-0456-z
20. Bhatia M, Girdhar A, Chandrakar B, Tiwari A (2013) Implicating nanoparticles as potential biodegradation enhancers: a review. *J Nanomed Nanotechol:* 4(175):2, http://dx.doi.org/10.4172/2157-7439.1000175
21. Gu J D (2003) Microbiological deterioration and degradation of synthetic polymeric materials: recent research advances. *International biodeteriorationand biodegradation*: *52*(2), 69-91 https://doi.org/10.1016/S0964-8305(02)00177-4
22. Satlewal A, Soni R, Zaidi M, Shouche Y, Goel R (2008) Comparative biodegradation of HDPE and LDPE using an indigenously developed microbial consortium. *J Microbiol Biotechnol:* 18(3):477–482
23. Restrepo-Florez J M, Bassi A, Thompson MR 2014 Microbial degradation and deterioration of polyethylene—a review. Int Biodeterior Biodegrad: 80:83–90.
24. Starnecker A, Menner M (1996) Assessment of biodegradability of plastics under stimulated composting conditions in a laboratory test system. *Int Biodeterior Biodegrad:* 37:85–92. https://doi.org/10.1016/0964-8305(95)00089-5
25. Gu J D, Ford TE, Mitton DB and Mitchell R (2000) Microbial corrosion of metals. *The Uhlig corrosion handbook,* 2nd edn. Wiley, New York: 915-27.
26. Vona I A, Costanza J R, Cantor H A and Robert W J (1965) Manufacture of plastics. *Wiley, New York*: *1*(66), 141-142. DOI: 10.1007/978-3-642-03287-5_1
27. Shah A A, Hasan F, Hameed A and Ahmed S (2008) Biological degradation of plastics: a comprehensive review. *Biotechnology advances*: *26*(3), 246-265. https://doi.org/10.1016/j.biotechadv.2007.12.005
28. Raziyafathima M, Praseetha P K and Rimal Isaac R S (2016) Microbial degradation of plastic waste: a review. *Chemical and Biological Sciences*: *4*, 231-42.

29. Kathiresan K (2003) Polythene and plastics-degrading microbes from the mangrove soil. *Revista de biologia tropical*: *51*(3-4), 629-633.
30. Soud S A (2019) Biodegradation of Polyethylene LDPE plastic waste using Locally Isolated *Streptomyces* sp. *Journal of Pharmaceutical Sciences and Research*: *11*(4), 1333-1339.
31. Singh B and Sharma N (2008) Mechanistic implications of plastic degradation. *Polymer degradation and stability*: *93*(3), 561-584. https://doi.org/10.1016/j.polymdegradstab.2007.11.008
32. Gonçalves F G, Brocco V F, Paes J B, Loiola P L and Lelis R C C (2014) Resistência de Painéisaglomerados de Acacia mangiumWilld. colados com ureia-formaldeído e taninos a organism oxsilófagos. *Floresta e Ambiente*: *21*(3), 409-415 https://doi.org/10.1590/2179-8087.059113
33. Gonçalves F G, Brocco V F, Paes J B, Loiola P L and Lelis R C C (2014) Resistência de Painéisaglomerados de Acacia mangiumWilld. colados com ureia-formaldeído e taninos a organism oxsilófagos. *Floresta e Ambiente*: *21*(3), 409-415 https://doi.org/10.1590/2179-8087.059113
34. Lopez, Yonny Martinez, GonÃ§alves, Fabricio Gomes, Paes, Juarez Benigno; Gustave, Donatian, Theodoro Nantet, Anna Clara, Sales, Tiago Jose (2020) Resistance of wood plastic composite produced by compression to termites Nasutitermes corniger (Motsch.) and Cryptotermes brevis (Walker). *International Biodeterioration and Biodegradation:* 152, 104998–. doi:10.1016/j.ibiod.2020.104998
35. Nuryawan A, Hutauruk N O, Purba E Y S, Masruchin N, Batubara R, Risnasari I and McKay D (2020) Properties of wood composite plastics made from predominant Low Density Polyethylene (LDPE) plastics and their degradability in nature. *PloS one*: *15*(8), e0236406. https://doi.org/10.1371/journal.pone.0236406
36. Yang X L, Wang S H, Gong Y and Yang Z G (2021) Effect of biological degradation by termites on the abnormal leakage of buried HDPE pipes. *Engineering Failure Analysis*: *124*, 105367. https://doi.org/10.1016/j.engfailanal.2021.105367
37. Lenz M, Lee C Y, Lacey M J, Yoshimura T and Tsunoda K(2011) The potential and limits of termites (Isoptera) as decomposers of waste paper products. *Journal of economic entomology*: *104*(1), 232-242 doi: 10.1603/ec10155
38. Thamil C T (2016) Isolation and identification of polyethylene biodegradation bacterial from the guts of plastic bags - eating damp wood termites. *Life Science Archives (LSA):* ISSN: 2454-1354 Page: 490 - 499
39. Kato, Kinya, Kozaki S, Imamura T and Sakuranaga M (2000) "*Pseudomonas cepacia* strain isolated from termite intestine that degrades trichloroethylene and furan compounds." *U.S Patent* 6,096,530.

40. Terzi E, Köse C, Büyüksarı Ü, Avcı E, Ayrılmış N and Kartal S N (2009) Evaluation of possible decay and termite resistance of particleboard containing waste tire rubber. *International Biodeterioration and Biodegradation*: *63*(6), 806-809.
41. Sharma R, Kaur R, Rana N, Poonia A, Rana D C and Attri S (2021) Termite's potential in solid waste management in Himachal Pradesh: A mini review. *Waste Management and Research*: 39(4), 546-554. https://doi.org/10.1177/0734242X20957394
42. Laessoe T (1994) Index ascomycetum 1. Xylariaceae. *Systema Ascomycetum*.
43. Rogers J D, Ju Y M and Lehmann J (2005) Some Xylaria species on termite nests. *Mycologia*: *97*(4), 914-923. DOI: 10.3852/mycologia.97.4.914
44. Mohan K (2011) Microbial deterioration and degradation of polymeric materials. *Journal of Biochemical Technology*: *2*(4), 210-215.
45. Kale S K, Deshmukh A G, Dudhare M S and Patil V B (2015) Microbial degradation of plastic: a review. *Journal of Biochemical Technology*: *6*(2), 952-961. DOI:10.31838/ijpr/2021.13.01.245
46. Tokiwa Y and Suzuki T (1977) Hydrolysis of polyesters by lipases. *Nature*: *270*(5632), 76-78. DOI: 10.1038/270076a0.
47. Tokiwa Y, Calabia B P, Ugwu C U and Aiba S (2009) Biodegradability of plastics. *International journal of molecular sciences*: *10*(9), 3722-3742. DOI: 10.3390/ijms10093722
48. Tokiwa Y and Calabia B P (2004) Review degradation of microbial polyesters. *Biotechnologyletters*: *26*(15), 1181-1189.https://doi.org/10.1023/B:BILE.0000036599.15302.e5
49. Henton, D. E, Gruber, P, Lunt, J, and Randall, J. (2005). Polylactic acid technology. *Natural fibers, biopolymers, and biocomposites*: *16*, 527-577.
50. Kijchavengkul T and Auras R (2008) Compostability of polymers. *Polymer International*: *57*(6), 793-804 https://doi.org/10.1002/pi.2420
51. Ho K L G, Pometto A L and Hinz P N (1999) Effects of temperature and relative humidity on polylactic acid plastic degradation. *Journal of environmental polymer degradation: 7*(2), 83-92 https://doi.org/10.1023/A:1021808317416

IMPACT OF COVID-19 PANDEMIC ON ENVIRONMENTAL, SOCIAL, AND GOVERNANCE INDEX IN INDIA

Manivannan Babu[1], A. Antony Lourdesraj[1], C. Hariharan[2], J. Gayathri[3], CS Karishma Butani[4], Chinnadurai Kathiravan[5]

[1] Bharathidasan School of Management, Bharathidasan University
[2] Nehru Institute of Technology
[3] Department of Commerce and Financial Studies, Bharathidasan University
[4] Institute of Commerce, Nirma University, Ahmedabad
[5]Assistant Professor, VIT Business School, Vellore Institute of Technology, Vellore
Corresponding author: drbabu@bdu.ac.in

Abstract: The current research investigates the impact of the COVID-19 Pandemic on Indian Sustainable Investment. For this purpose, the study selected three sustainable indices, namely, BSE Greenex, BSE Carbonex, and BSE 100 ESG index of Bombay Stock Exchange of India. The Descriptive Statistics, ADF Test and GARCH (1,1) Model were used to found that all the sample sustainable indices were gained positive return with the high volatility throughout the COVID-19 Pandemic and the results confirmed that COVID-19 influenced the Indian sustainable investments.

1. Introduction

Stock exchanges across the world, according to the efficient market theory, pay great attention to market news. Almost every financial markets responded to the existence of any negative news in the markets, regardless of whether such crises are directly associated to the financial sector or even have any indirect impact over it. Recent research indicates that, aside from the global financial crisis, the breakout of infectious illnesses which including severe acute respiratory syndrome (SARS) and H1N1 enhanced stock market volatility (Mun and Brooks, 2012; Peckham, 2013; Jin and An, 2016).The whole earth has just been afflicted by the highly communicable and dangerous virus COVID19. Not only did the COVID-19 epidemic impair people's health, but it also had a detrimental influence on financial markets and enterprises. The government's enforcement of curfews, lockdowns, travel bans, and the closure of industries, retail stores, trades, and enterprises has shaken all economies. Global financial markets have become very turbulent as a result of the COVID-19 epidemic since December 2019. Yousef (2020) discovered that COVID-19 has a significant impact on conditional volatility for all main US stock market indexes. Similar discoveries are predicted in the Indian financial industry. However, with the mindset of a risk-averse investor, the current study seeks to investigate an investing path that may bring some relief during this stressful period. Researchers (Tripathi and Bhandari, 2012; Aggarwal, 2014) showed that green and sustainable equities were fairly constant and outperformed other portfolios throughout the volatile period of the financial crisis. Thepurpose of this research is to determine if it is beneficial to invest in ESG stocks during the volatility induced by the emergence of coronavirus. The research looks at the impact of stock volatility, which is a huge concern for all investors, traders, company owners, and regulators. The findings are also intended to provide factual reasons for why a company should indicate ESG commitment. The next part examines the available literature before moving on to the research approach used for analysis. The fourth portion investigates the findings, and the last section summarizes the study.

2. Literature Review

During the last decade, the notion of "socially responsible investment," that is an investment built on ESG parameters, has gained a lot of popularity. The approach asks for the formation of new, sustainable products. Massive investment in these kinds of funds is supporting the development of green and sustainable stocks in the market. Over a decade, the number of sustainable funds available to US investors almost fourfold (Hale, 2021).Tripathi and Bhandari (2015), as well as Kaur (2018), investigated the performance of Sustainable indexes like the ESGindex and the Greenex index. The study made a comparison of these indices to the performance of two traditional Indian indices, the Nifty and Sensex. The data demonstrated that, as compared to traditional indexes, sustainability indices have positive returns throughout the crisis era. Malik and Yadav (2020) investigated the profitability and volatility of sustainable indices such as Greenex, Carbonex, and ESG index. The GARCH model results revealed that the variance of all these indices is a function of its prior history. However, Onali (2020) found that the cases and fatalities of COVID-19 had no effect on the returns of US financial markets. Few studies found that infectious disease outbreaks have an impact on key stock markets and their worldwide returns (Senol and Zeren, 2020). Investing in sustainable indices, on the other hand, provides enough financial returns throughout COVID-19 (Sharma et al., 2021; Omura et al., 2021). Thus, several researchers have dissected various aspects of the individual stock market. Some studies examined the volatility, risk, and return characteristics of sustainable enterprises, while others performed research to assess the impact of various shocks. However, research evaluating COVID-19's impact on sustainable indicators. market capitalization-based indices. In thiscontext, the current study seeks to fill a research gap. The study is to test the null hypothesis that COVID-19 has no substantial influence on the volatility of chosen index returns (i.e. BSE Greenex, Carbonex, Large-Cap, Mid-Cap and Small-Cap).

3. THE STUDY'S OBJECTIVES

The purpose of this research is to examine the impact of the GOVID-19 pandemic on the volatility of Indian sustainability indexes. The current study additionally established the following analysis objectives.

To investigate the normalcy and stationarity of daily pricing for Indian sustainable indicators during the GOVID-19 Pandemic.

To investigate the daily volatility of Indian sustainable indexes during the GOVID-19

Pandemic.

4. THE STUDY'S NULL HYPOTHESES

NH01: During the GOVID-19 Pandemic, there is no fluctuation in the daily returns of Indian sustainable indexes.

5. METHODOLOGY

The BSE of India has successfully launched three sustainable indices: BSE Greenex, BSE Carbonex and BSE 100 ESG Index. Daily observations for all indexes were obtained from the BSE's official website beginning on 1 April 2016 and ending on 31 March 2022.To begin, logarithmic daily returns were estimated using the following formula:

$$R_t = \ln \frac{P_t}{P_{t-1}}$$

where R denotes the return during the 't' time period Pt denotes the price of the stock at the end of the time period; Pt-1 denotes the price of the stock at the start of the time period; and ln denotes the natural log.

For testing the study's hypotheses, the researchers employed the following statistical methods.

1. Descriptive Statistics
2. ADF Test
3. GARCH (1, 1) Model.

6. FINDINGS AND EXPLANATION

Table- 1 Descriptive statistics for daily returns of the Indian sustainable indices duringCOVID – 19 Pandemic.

	BSE 100 ESG Index	BSE Greenex	BSE Carbonex
Mean	0.00112	0.00129	0.00201
Std. Dev.	0.01138	0.01198	0.02317
Skewness	0.02054	0.19121	0.52588
Kurtosis	4.23063	4.30453	5.19263
Jarque-Bera	11.879	14.478	46.325

Table- 2 Results of ADF Test for the Indian sustainable indices during COVID – 19 Pandemic.

		BSE 100 ESG Index	BSE Greenex	BSE Carbonex
ADF Test		-11.7112	-11.5705	-13.4592
Test critical values:	**1%**	-3.46539	-3.46539	-3.46539
	5%	-2.87684	-2.87684	-2.87684
	10%	-2.57501	-2.57501	-2.57501

b) Results Volatility Test

Table-3 displays the GARCH Model's results for daily returns of sample Indian sustainable indices during the GOVID-19 Pandemic period. The alpha and GARCH values revealed that the sample Indian sustainability indices' daily returns were volatile during the GOVID-19 Pandemic period. When price variations are high, investors can often earn higher returns on their investments. As a result, the Null Hypothesis NH01, "During the GOVID-19 Pandemic, there is no fluctuation in the daily returns of Indian sustainable indexes" was rejected.

Table- 3
the GARCH Model's results for daily returnsof the Indian sustainable indices during COVID – 19 Pandemic.

	BSE 100 ESG Index	BSE Greenex	BSE Carbonex
α	-0.02516	-0.02955	-0.0149
β	1.008465	1.017782	1.026233
α+β	0.983304	0.98823	1.011332

α - ARCH (-1) ; β - GARCH(-1)

CONCLUSION

Financial markets are extremely susceptible to market shocks, regardless of whether they are directly related to the financial world (e.g., the financial crisis) or have any indirect impact on it (e.g., outbreak of contagious diseases). Recognizing the importance of sustainable supplies, the current article seeks to determine how resistant these stocks are to the current pandemic. Thus the research study utilize GARCH model to examine the volatility of the sample indexes throughout the GOVID-19 Pandemic. The results revealed that the daily returns of sample indices were volatile, throughout the GOVID-19 Pandemic. The COVID-19 impact was observed by the GARCH model over BSE 100 ESG Index, Greenex, and Carbonex. The current study's findings may assist investors in determining the risk exposure of their investments in various indexes. Investors may use the findings to help shape their investing plan and better manage their portfolio. Furthermore, it provides factual support for the necessity to commit to the environment and society. Because the results are centred on the Indian capital market, future research may include an orientation to the worldwide capital market. The influence of COVID- 19 on the risk and return characteristics of sustainable indices in comparison to no sustainable indices may be investigated. It would also be interesting to investigate any volatility spillover effects of these indexes.

REFERENCES

1. Aggarwal, K. (2014), "What determines the investing in green companies in India: an empirical analysis", *Aspirare*, Vol.1, pp. 108-114.
2. Estrada, M.A. and Lee, M. (2020), "How COVID-19 can affect the worldwide stock markets?",Availableat SSRN 3591698.
3. Hale, J. (2021), "U.S. sustainable funds continued to break records in 2020", available at: www.morningstar.com/articles/1026261/us-sustainable-funds-continued-to-break- records-in-2020(accessed 12 December 2021).
4. He, Q., Liu, J., Wang, S. and Yu, J. (2020), "The impact of COVID-19 on stock markets",Economic and Political Studies, pp. 1-14.
5. Jin, X. and An, X. (2016) "Global financial crisis and emerging stock market contagion: a volatilityimpulse response function approach", Research in International Business and Finance, Vol. 36,pp. 179-195.
6. Kaur, R. (2018), "BSE Greenex: adding greenness to Indian stock market", Journal of Management,Vol. 5 No. 3, pp. 24-31.
7. Lacurci, G. (2021), "Money invested in ESG funds more than doubles in a year", Published onFebruary11, 2021, available at: www.cnbc.com/2021/02/11/sustainable- investment-funds-morethandoubled-in-2020-.html (accessed 12 December 2021).
8. Liu, H., Manzoor, A., Wang, C., Zhang, L. and Manzoor, Z. (2020), "The COVID-19 outbreakand affectedcountries stock markets response", International Journal of Environmental Research and Public Health, Vol. 17 No. 8, p. 2800.

9. Malik, C. and Yadav, S. (2020), "Forecasting and asymmetric volatility modelling of sustainability indexes in India", Corporate Governance and Sustainability Review, Vol. 4 No. 1, pp. 56-64.
10. Morales, L. and Andreosso-O"Callaghan, B. (2020), "Covid19: global stock markets BlackSwan", Critical Letters in Economics & Finance, Vol. 1 No. 1, pp. 1-14.
11. Mun, M Brooks, R. (2012), "The roles of news and volatility in stock market correlations during the global financial crisis", Emerging markets review, Vol. 13 No.1, pp.1-7.
12. Omura, A., Roca, E. and Nakai, M. (2021), "Does responsible investing pay during economic downturns: Evidence from the COVID-19 pandemic", Finance Research Letters, Vol. 42, 101914.
13. Onali, E. (2020),"Covid-19 and stock market volatility", Available at SSRN 3571453
14. Ozili, P.K. and Arun, T. (2020), "Spillover of COVID-19: Impact on the global economy"
15. Peckham, R. (2013), "Economies of contagion: financial crisis and pandemic", Economy and Society, Vol. 42 No. 2, pp. 226-248.
16. Ramelli, S. and Wagner, A.F. (2020), "Feverish stock price reactions to covid-19",
17. Senol, Z. and Zeren, F. (2020), "Coronavirus (COVID-19) and stock markets: The effects of thepandemicon the global economy", Eurasian Journal of Researches in Social and Economics (EJRSE), Vol. 7No. 4, pp. 1-16.
18. Sharma, G.D., Tiwari, A.K., Talan, G. and Jain, M. (2021), "Revisiting the sustainable versus conventional investment dilemma in COVID-19 times", Energy Policy, Vol. 156,p. 112467.
19. Tripathi, V. and Bhandari, V. (2012), "Green is good in Indian stock market", Colombo Business Journal,Vol. 3 No. 2, pp. 27-45.
20. Tripathi, V. and Bhandari, V. (2015), "Socially responsible stocks: a boon for investors in India",Journal of Advances in Management Research, Vol. 12 No. 2, pp. 209-225.
21. Yousef, I. (2020), "Spill over of COVID-19: impact on stock market volatility", International Journal of Psychosocial Rehabilitation, Vol. 24 No. 6, pp. 18069-18081.
22. Zhang, D., Hu, M. and Ji, Q. (2020), "Financial markets under the global pandemic of COVID19" Finance Research Letters, 101528.

AUTHOR INDEX